"VICTORIA RIFLES,"
(ENROLLED AS A VOLUNTEER REGIMENT, 1853.)
Lieutenant-Colonel Commandant, His Grace the Duke of Wellington,
&c. &c. &c.

RIFLE VOLUNTEERS
HOW TO ORGANIZE AND DRILL THEM

by
HANS BUSK

1859

The Naval & Military Press Ltd

Published by the
The Naval & Military Press
in association with the Royal Armouries

Unit 10 Ridgewood Industrial Park,
Uckfield, East Sussex, TN22 5QE
Tel: +44 (0) 1825 749494
Fax: +44 (0) 1825 765701

MILITARY HISTORY AT YOUR FINGERTIPS
www.naval-military-press.com

ONLINE GENEALOGY RESEARCH
www.military-genealogy.com

ONLINE MILITARY CARTOGRAPHY
www.militarymaproom.com

ROYAL ARMOURIES

The Library & Archives Department at the Royal Armouries Museum, Leeds, specialises in the history and development of armour and weapons from earliest times to the present day. Material relating to the development of artillery and modern fortifications is held at the Royal Armouries Museum, Fort Nelson.

For further information contact:
Royal Armouries Museum, Library, Armouries Drive,
Leeds, West Yorkshire LS10 1LT
Royal Armouries, Library, Fort Nelson, Down End Road, Fareham PO17 6AN

Or visit the Museum's website at
www.armouries.org.uk

In reprinting in facsimile from the original, any imperfections are inevitably reproduced and the quality may fall short of modern type and cartographic standards.

PREFATORY OBSERVATIONS.

On the 3rd of May last the Right Hon. T. Sotheron Estcourt, Secretary of State for the Home Department, in a speech delivered at Devizes, made the following official announcement:—" I take this public opportunity of saying, that if in this county and in any others there are any gallant spirits ready to enrol themselves in Rifle Corps, or similar Volunteer bodies, the Government will be glad to receive the assurance of their willingness to do so; and as far as we are concerned, and as Parliament will sanction the measure, we shall be glad to afford them countenance and good will."

Now, if this means anything at all, it implies a resolution, on the part of the Ministry, to bring forward a practical scheme for properly organizing the defences of the country. At page 40 will be found one, which has been well considered; and though some few alterations might perhaps be advantageously introduced, I print it as it was submitted

iv PREFATORY OBSERVATIONS.

to me, in preference to making or suggesting such emendations at present.*

We shall shortly know upon what footing Volunteer forces will be placed by the new Government measure. A bare reference to the Act of 1804, without a careful adaptation of its provisions to modern exigencies, would be simply puerile, and would evince an intention of trifling with a large and important class of the community.

In the meanwhile, there is no reason why we should not bestir ourselves. No time should be lost in seconding the efforts of the Executive to add to the security of the realm. We can all do something; either by the promotion of the formation of Rifle Clubs and Rifle Corps, by collecting subscriptions to cover the preliminary expenses, by instructing others in the art of rifle-shooting, or by setting apart a portion of our time to attaining proficiency ourselves.

Remember, that every man who has the use of his eyes may in a few weeks make himself something of a shot, even when so situated that he has not access to a practising ground. I will explain how. All the requisite apparatus is a rifle, the

* The proposed plan has the great recommendation of economy. It not only calls for no present outlay, but it points out a certainty of retrenchment, by a diminution of the heavy disbursements caused by a large standing army, while the greater part of the cost to be incurred on behalf of Volunteers would be transferred to a period some quarter of a century hence.

trigger of which does not require a pull of more than 3lbs., a sand-bag, and a box of caps.

1. Take a sheet of cartridge-paper, fix a large red wafer on the centre, and hang it against the wall of your room. Put a chair upon the table, and lay the sand-bag in a heap on the seat of the chair. Now lay your rifle upon the sand-bag, so that it will remain in any position in which you place it, and then looking through the back sight at the fore sight, let the apex of the fore sight cover the wafer. Habituate yourself to this simple practice, and when you have learnt the first lesson,

2. Lay a piece of stout leather upon the nipple of the gun, as it rests upon the sand-bag, pull the trigger, and observe, when you have done so, how much the sight is pulled away from the wafer. When it ceases to show any deflection,

3. You may dispense with the chair and sand-bag, carry the rifle to your shoulder, and aiming at the wafer, snap a few caps till you can do so without blinking or starting at the explosion; to ascertain whether you have done so or not, always keep the rifle to your shoulder for a second or two after each discharge, and note particularly the aberration of the muzzle sight.

4. Now, standing at about $2\frac{1}{2}$ or 3 yards from a lighted candle, aim at the base of the flame, and snap a few caps at it till you can succeed in putting it out eight times out of ten.

5. The next lesson will be at an iron target with

bullet, and about a third of the usual charge of powder. Mark a bull's-eye about 2 ft. in diameter, and at a distance of 20 yards aim deliberately and fire. When you feel confident of doing this every time, step back a couple of yards at each shot, till you get to a hundred yards. Now use the proper quantity of powder, and practice daily at that distance for an hour or so, for some weeks. Then, still practising at 100 yards, reduce the diameter of your bull's-eye, till it is not more than 10 inches wide; and, proceeding as before, retire till you get to 200 yards.

More precise instructions on the subject will be found in the Fourth Edition of the "Rifle." When the tyro has patiently mastered these, the probabilities are, that he will be better qualified to join a Rifle Club than nine-tenths of the usual run of candidates.

Nothing will contribute more to the rapid attainment of proficiency than keeping a record of each day's practice; a simple and inexpensive mode of doing this is afforded by the "RIFLE TARGET REGISTER," published by Messrs. Routledge. It consists of a series of diagrams, each representing a regulation target, and superficially divided into squares. The precise situation of each shot is jotted down, either with a pin or pencil, by a bystander or by the shooter himself.

CONTENTS.

	PAGE
INTRODUCTION	1
ON THE ORGANIZATION OF VOLUNTEER CORPS	8
ON VOLUNTEER REGIMENTS	28
ORGANIZATION OF VOLUNTEER CORPS AS DISTINCT FROM CLUBS	54
MANUAL AND PLATOON EXERCISES	63
THE NEW MANUAL EXERCISE FOR RIFLEMEN	66
THE PLATOON EXERCISE	75
MUSKETRY INSTRUCTION	95
LIGHT INFANTRY	97
MISCELLANEOUS MOVEMENTS	109
ABSTRACT OF LAWS RELATING TO VOLUNTEER CORPS	113

RIFLE VOLUNTEERS.

INTRODUCTION.

I HAD scarcely anticipated that the observations lately addressed by me anonymously, to the British public on the subject of VOLUNTEER RIFLE CORPS, in the columns of the *Times*, would so soon have been followed by such highly gratifying results.

The spirited correspondence which ensued through that medium, not to mention the numberless letters addressed to me by every post soliciting information as to the means of organizing rifle clubs, &c., proved that the matter was one regarded by the community as of vital moment; and the important official announcement since issued shows that the Government have resolved once more to put in force a measure which has been lying dormant in the statute-book for more than half a century. The fiat of her Majesty has gone forth, and

THE VOLUNTEERS ARE TO BE CALLED OUT.

The announcement has everywhere been received with acclamation, if not with enthusiasm. That noble body, the University of Cambridge, was the first to respond to the call to arms: within a few days nearly a thousand pounds was there subscribed, and several hundred names are already enrolled as a club, preparatory to the formation of a Corps. Oxford was not behind-hand, but the authorities in that august seat of learning, seem scarcely to have exhibited the loyalty, promptitude, and energy displayed at Cambridge. Be that, however, as it may, both are now actively at work; and we shall shortly have, from the very *élite* of the classes residing in those two cities, two of the finest corps of Volunteers that the country could produce.

The large towns throughout the Empire are bestirring themselves; and, as I predicted nearly three years ago, when first advocating the subject, the array of Riflemen now coming forward, will well enable us to treat with scorn the bluster of the most quarrelsome of French swaggerers; and we may shortly feel perfectly assured upon the subject of our capacity to resist invasion.

With even one hundred thousand skilled marksmen scattered over the land, capable of mustering rapidly, and of assembling in force at any required point, we should hear no more insulting talk across the Channel about "our existing by French sufferance" — "holding our country only until it suited our neighbours to annex it;" together with much other similar insolence, such as has for years past, and more especially during the last few months, interlarded most of the inflammatory publications daily emanating from the Parisian press.

But we have still work—and important work— before us. We have men enough, willing, eager, "ready," as the old Duke said, the gallant Rifle Brigade always were, "to do anything and go anywhere;" but at the present moment the majority of them have for the most part everything to learn, they are but recruits, confident, no doubt, and zealous in a good cause, but still unskilled. How, then, are they to be converted into SOLDIERS?

I propose accordingly to point out the simplest way in which this is to be done; and as the majority of my numerous correspondents seem to consider that I have taken the initiative in the great movement now going on throughout the land—although I do not adopt that view, nor wish to arrogate to myself so high an honour—I shall endeavour presently to show, how the "movement" may best be turned to account.

It is difficult to understand the supineness of our authorities hitherto in neglecting to take advantage of the readiness which has so often been evinced by the finest portion of the population to organize themselves for her defence. A very slight encouragement—a few trifling privileges—an unimportant modification of the Volunteer Act of 1804—and in a few weeks, without the outlay of a farthing, we might long since have had enrolled in each of the principal counties, where a force of this description would be of the greatest avail, regiments 800 or 1000 strong. In 1798, when the danger of invasion was supposed imminent, 60,000 volunteers came forward in a few days, and the Bishop of Winchester (as appears from the *Times* of the 12th April of that year) authorized the whole of the clergy of Hampshire, and especially of the Isle of Wight, to take up arms and to do whatever they might think best for the service of their country. It is not worth while to pause to consider what impediment the 60,000 un-drilled Volunteers, hastily collected, would have offered to the legions of Bonaparte, flushed with victory, had they once effected a landing; but there cannot be the smallest doubt that the same number properly disciplined and accustomed to cooperate with, what for the sake of distinction we may term "mercenaries," would not only afford most valuable aid, but would excite salutary emulation in the breasts of their brethren of the Line.

"We know," said the *Times*, very justly, in May, 1859; "that if the hour of danger should come, Volunteers would rush forward by hundreds of thousands; but it is to obviate such a panic that we would ask for the speedy restoration of what may be called the old national force of the country. It is not Line regiments—it is not Militia that we want—but men of ordinary occupations, trained by a certain amount of drill to support the regular armed force either in the field or the fortress. There is plenty of wealth and plenty of leisure for the formation of such corps, and the metropolis alone could furnish a contingent strong enough to garrison any two naval ports of the kingdom."

But we are in all probability about to witness a considerable change; the voice of the country has been already heard, and it will speak out still more loudly ere long, and noodle-dom and fogey-ism must shelve their silly prejudices, as they often have many more useful things ere now.

The "United Service Magazine," the organ both of the Navy and Army, and therefore representing no doubt the opinions of the most enlightened officers in both services, observes—"Why have we not plenty of Rifle Regiments—why are not our men as famous with the rifle as our forefathers with the bow?"

Far wiser indeed were our ancestors; while other countries only prepared for war when it was imminent, England, with superior vigilance and intelligence, adopted strenuous measures to maintain her pre-eminence in the use of her then favourite arm.

At one time, by law, boys at seven years of age were compelled to practise its use, and an adult was not allowed, under a penalty, to shoot at a target placed at a less distance than 220 yards. A restraint was put upon other games and sports, lest they should interfere with archery. By a stringent law, every Englishman was bound to provide himself with a bow and arrows. The old yews that now adorn many of our village grave-yards were planted with no other object than to supply the requisite materials for the use of our matchless archers.

Every parish throughout England was obliged to maintain the necessary "butt," and the afternoons of Sundays and holidays were set apart for healthful exercise.

In the thirty-third year of our Eighth Henry, when fire-arms were coming into general use, an enactment was passed for the encouragement of the use of hand-guns, which were to be " of the length of one whole yard, and not under," the reason given being, that those who used them, "and every of them, by the exercise thereof, might the better aid and assist to the defence of this realm, when need shall require."

What the bow was of yore, let the rifle now become; let it be the pride of every one entitled to call himself an Englishman, to attain proficiency in the use of a weapon the most perfect and the most formidable, that human ingenuity has yet devised.

The following observations, by one of the ablest writers of the day, point out so eloquently the folly of which past Governments have been guilty, in neglecting the best and most natural defence of the country, and the importance of at once following an opposite course, that no apology is needed for introducing here, sentiments to which too much publicity cannot be accorded. They deserve to be emblazoned in letters of gold, and to be enshrined over every English hearth :—

"A deep disgrace it is, to those who in later times have had the direction of our Army, that the public spirit of the country has been uniformly checked and its ardour cooled by military personages. Our older officers, those to whose obstinacy and incapacity we owe the disasters of the Crimea, and indirectly the complications of the present hour, have constantly discouraged the formation of those corps to which countries like Prussia and the United States look for their safety. To keep 'civilians' apart, as entirely distinct from military men, to have the regular army everything and all the rest of the nation nothing, has been the policy of the old fogies of the Horse Guards, and it is of a piece with all that we know of pre-Crimean administration. 'Volunteers, sir!' would exclaim a plethoric Colonel; 'all nonsense, sir; it takes three years to make a soldier—not an hour less. Volunteers in war are not of the slightest use—only get in the way, sir.' This has been the enlightened criticism of the Regulars, for many a year past, and it is likely to prevent all attempts at creating a cheap and efficient force for service in these islands, unless the popular voice be loudly heard. The only notion of patriotism current in certain quarters is the payment of taxes for new regiments. More money, and still more money, and ever more money, is the cry. But there must be a limit both to taxes and standing armies. For the purposes of national defence we have a mine never as yet explored, and which we may predict to contain immense resources. It is the spirit of our young men throughout the country, and their willingness to form themselves into bodies for the support of the regular troops in case of need. Educated men are not like ploughboys, and there cannot be a doubt that a few days' drill and rifle practice, in the month, would shortly give us a hundred thousand men perfectly fit for the ordinary duties of a soldier. Commanded by officers who have served in the regular army, accustomed to the use of a perfect weapon, what strength would they not impart to this country on the eve of a war! What a gain would it not be, in a moment of peril, if Portsmouth, and Plymouth, and Chatham required only a few hundred regular Artillerymen, and if all the rest of the duty could be done by the Volunteer levies of London or Liverpool, or half-a-dozen agricultural counties, while the whole strength of the Line and Militia was ready to meet an invader in Kent or Sussex! That such a force is indispensable, and that it is easy to establish and maintain, we firmly believe, and a grave responsibility will rest on the Government which neglects to encourage the spirit which exists among the people."

SECTION I.

ON THE ORGANIZATION OF VOLUNTEER CORPS.

It is a great mistake to suppose that all that is requisite to constitute an useful and efficient Rifleman is to put on a dark tunic, and blaze away for a few hours daily at a target. "Ready and dexterous," like the archers of old, you may be; but there is a good deal more knowledge to be acquired before you can be considered a soldier, and before you will be fit to co-operate with troops in the field.

You may be a very creditable member of a Rifle Club, and yet, were you to be called out, even for an ordinary field-day, and required to act in concert with trained soldiers, you would not only cut a sorry figure, but would most certainly soon become bewildered yourself, and cause embarrassment to others.

Now Volunteers, to be of any use at all, must at least be tolerably proficient in Light Infantry drill; they must be accustomed to act and move together, understand the various words of command and bugle calls, and obey their officers with smartness and alacrity. They are not expected, nor would they ever be required, to exhibit the faultless precision of the Guards; but if they have not sufficient memory to remember a few general principles, nor patience to practise some very simple movements, they had far better leave the rifle to other hands, and not venture to don an uniform.

The whole of the Light Infantry drill has of late

been revised and simplified; much that had long been deemed superfluous has been dispensed with; almost all existing manuals have consequently been rendered obsolete, and, if now adopted by any corps, would prove worse than useless.

In the instructions given in this volume, I have endeavoured to follow, as closely as I considered requisite, the drill now adopted and taught throughout her Majesty's service, only simplifying it, where that could be done, without impairing its utility. But, before proceeding to the more important branch of our subject, it may be advisable to offer a few observations on

RIFLE CLUBS.

Rifle Clubs, or mere associations for the purpose of target practice, without any immediate military object, and, as such, distinct from Rifle Corps or Regiments, may be established at any place or under any regulations which their members may think proper to adopt. A military corps, however, must in the first instance necessarily be enrolled under the provisions of the Volunteer Act (44 Geo. III., cap. 54), an abstract of which, together with all the subsequent statutes in any way affecting it, will be found at the end of the volume.

Rifle Clubs, nevertheless, as auxiliary to military corps, cannot but be deemed to be of considerable national importance. The first thing to be done in organizing them, is to determine the amount of the subscription; the site for the ground, the days and hours for practice, the amount and value of the prizes to be periodically awarded, and other similar matters of detail.

It will be found advisable to institute, in each club, three classes. All members joining to enter the third or lowest class, and to be gradually promoted, as they prove their proficiency in shooting, till they attain the first, when, and not till then, they may be eligible to be drafted into a Volunteer Regiment, which would thus at least consist of picked shots. In most counties one such regiment will be found sufficient, but there may be an unlimited number of clubs.

ORGANIZATION.

We will now suppose the preliminary difficulties overcome, and that some fifty or a hundred members have agreed to establish an association in their district to practise rifle shooting.

In order to provide funds to cover the unavoidable expenses, they should pay an entrance fee say of a pound, with an annual subscription of a pound, claimable yearly in advance.

Each member must of course provide himself with a rifle and with ammunition. It will be an advantage if the gauge of these rifles be similar, but this is not strictly essential.

One member should undertake to give his services for a twelvemonth gratuitously, as secretary, the functions of that somewhat important and occasionally onerous office to be assumed the following year by another member, and so on in rotation.

The duty of the secretary will be, to convene meetings, enforce the rules, collect the subscriptions and fines, make out from time to time lists of members, issue circulars, &c. The success and prosperity of a club will be found to depend greatly on the tact, intelligence, and energy of this officer.

Where the funds of the society will warrant the disbursement, the secretary should receive a moderate salary, and there should be besides a marker constantly upon the ground, if not a serjeant armourer, to clean the rifles and keep them in order. An instructor of musketry, recommended by General Hay (the head of the Government establishment at Hythe), would be a valuable and most important adjunct, if the club can bear the expense.

With regard to the ground, if the grant of a piece of waste land cannot be obtained, a narrow strip must be rented in the most convenient locality. It should be level; if covered with sward, so much the better, and care should be taken that there are no houses, roads, nor footpaths in the direction of the butt or target, and that the practice ground be fenced in. Where practicable, it is most desirable that the target should stand at the foot of a hill, bank, or mound, or in a gravel pit or quarry.

With respect to the range, five hundred yards is the greatest length that need be required under any circumstances, two hundred the least that will suffice, though beginners need not practice beyond one hundred. Where danger is likely to be apprehended from wild shooting, stray shots, &c., it may be much diminished by the erection, across the practice-ground, of two or three light iron arches, through which the shooters will necessarily have to aim.

Where no natural bank can be rendered available as a back-ground for the target, an artificial mound should be formed of earth, sand, clay, or turf; it ought not to be less than ten or twelve feet in thickness, and should extend twenty-five or thirty feet above the target, flanking it for a similar extent on either side.

A brick wall, thirty feet high and sixty feet wide, protected with wrought-iron plates, two inches thick, at that part immediately behind the target, will prevent almost the possibility of accident arising from the most careless shooting, especially if aim has to be taken through a hoop or arch, as already suggested.

The target itself, should be of cast iron, six feet high and two feet wide, and from one and a-half to two inches in thickness. The form and appearance of a regulation target is shown at page 19.

For practice up to two hundred yards, a circular bull's eye, painted black or red, should be marked, increasing in diameter to ten inches, for the four hundred yard range.

A space of about twenty feet in length by ten feet wide in front of the target should be kept bare, and well beaten and rolled for the convenience of recovering the bullets. To facilitate this, too, the target should not be perpendicular, but should slope slightly forward, by which means the fragments of the bullets striking it will be thrown to the ground, instead of being scattered in all directions and lost. Such indeed is the force of the explosion of modern rifles, owing to the abolition of windage, that at four hundred, and even five hundred yards, the leaden missiles break against the target into minute splinters, which fly with great force to a distance often of thirty or forty yards, and in time do much damage to the face of the solid brickwork in the rear and at either side.

This should not be forgotten, as serious accidents have often occurred to by-standers, who might have been supposed to be incurring no risk whatever. The marker should be well protected, either by a mound on one side of the target, or by what is far better, a hut, thickly plated with two layers of wrought iron, each not less than a quarter of an inch thick, with inch oak or elm boards between the two plates.

To obviate the necessity of his having to leave his

ORGANIZATION OF VOLUNTEER CORPS. 13

cover after each shot in order to point out its precise position, he should be provided with three small flags, fixed to long rods or wands—one white, signifying "one," and indicating when shown that the shot last fired, called an "outer," and counting one, struck one of the sixteen squares either above or below the sixteen centre squares; a dark blue flag to denote that the space termed the "centre," between the bull's eye and the outer circle, was struck, counting two; while a red and white flag should be shown after every bull's eye, three being counted for each such hit.

When a cast-iron target is not to be had, a substitute may be made by means of a wrought-iron frame, constructed somewhat upon the principle of those used by ladies for embroidery, and called a "tambour" frame. Over this a covering of stout cotton should be stretched, and upon that, a coating of "cartoon" or "lining-paper." The marker should in this case be provided with some paste and a supply of pieces of paper about two inches square. These he should apply over the perforations as they are made. A target thus constructed and repaired will last much longer than might at first be expected. It is scarcely necessary to add that when a target like this is used, there must be a solid mound of sand, or a bank behind, to stop the bullets. Remember that a bullet from an Enfield rifle, with the regulation charge of powder, will kill a man at 2000 yards, or even at a still greater distance.

The greatest possible care should therefore always be taken in handling so deadly an arm. Something more than a mere nominal fine should be imposed upon any member of the club who may be observed with his rifle at full cock, until he is actually prepared to fire, and the marker has retreated into his box. It is surprising how much the shooting will be found to improve if a fine (even of sixpence) be imposed every time a bullet strikes the brickwork, at such times as the practice does not exceed 200 yards.

Where the numbers and importance of the club render more formal rules necessary than those already alluded to, some such code as the following may be adopted. In that case they should be printed, a copy supplied to each member upon his election, and he should be required to subscribe his name to them in a book kept for the purpose. It may be added, that these rules have already worked well in more than one club where they have been adopted, and have not, after a lapse of years, been found to need either alteration or amendment.

RIFLE CLUB RULES.

I. A General Meeting shall be held annually, on or about the second Wednesday in July, of which at least seven clear days' notice in writing shall be given.

II. The affairs of the Club shall be under the control of a Council, consisting, *ex officio,* of the treasurer and secretary for the time being, and also of seven members, to be elected annually by ballot at the general meeting. Five to form a quorum.

III. All candidates for election to the Council shall be proposed by a member in writing, and sent to the secretary at least seven days before the day of election, with the consent of such candidate. Ballot lists shall be provided at the time of election.

IV. The Council shall meet at least once a month, and keep minutes of its proceedings. It shall have power to elect new members, to fill up any vacancy in the Council, and to appoint the treasurer and secretary, and also to make Bye-laws for the regulation of all matters committed to its charge.

V. Every candidate for admission to the Club shall be recommended by two members, who shall deliver, in writing, to the secretary, the name, place of abode, and profession or occupation of the candidate.

VI. Every gentleman, on his election, shall cause to be paid to the secretary one guinea as an entrance fee, and two guineas as his subscription for the current year. Members elected after the 1st of November, and before the month of May following, shall pay a subscription of one guinea only for that year.

VII. Every member shall, on the 24th of May in every year, cause to be paid to the secretary two guineas as his subscription for the ensuing year.

VIII. The Council shall have the power of erasing from the list of members any person whose subscription is more than one month in arrear.

IX. Any gentleman wishing to discontinue his membership of the Club, shall give notice of the same to the Secretary, in writing; but all arrears of subscriptions or fines due by him must be paid.

X. At any meeting of the Council or general meeting, the chairman for the time being shall not be entitled to vote, except on the occasion of an equality of votes, when he shall have a casting vote.

XI. At the annual general meeting, two auditors shall be elected.

XII. The accounts of the Society shall be made up to the 24th May in each year, duly audited, and laid before the next annual general meeting.

XIII. The property of the Society shall be vested in three trustees, to be elected by a general meeting of the members, such trustees to remain in office till death or resignation, or until removed by the vote of a general meeting.

XIV. Any member breaking any of the Laws or Bye-Laws of the Club, or otherwise misconducting himself, shall be dealt with according to the discretion of the Council, with the right, however, of appeal to a general meeting.

Efforts have been made on various occasions to establish similar associations composed of subscribing members, and of free or poorer members, the latter admitted by ballot, and provided at the expense of the former, with arms, accoutrements, &c., but the principle has not been found to answer the excellent intentions of its originators. In fact, from reasons that will be tolerably obvious, no association of the kind can be said to possess the elements of success, in which any attempt may be made to amalgamate different classes. The labouring man feels shy and embarrassed when admitted temporarily to a pretended equality with his superiors, who, on the other hand, feel a degree of repugnance to wearing the same uniform as their humble dependent, and admitting him to unwonted familiarity.

UNIFORM.

The uniform, if indeed any such distinction be deemed necessary, for a Rifle-club, cannot be too plain and simple. Ease to the wearer, lightness, and moderate warmth are the main requisites. As it is only required to be worn on practising days, and without regard to actual service in the field, the colour is of little consequence; a tunic or jerkin of dark grey, or of heather-coloured cloth, with broad-brimmed soft felt hat, such as was worn in the time

ORGANIZATION OF VOLUNTEER CORPS. 17

of Charles I., a leathern belt and capacious pouch, thoroughly waterproof, are all the equipment needed.

One pocket should contain a spare nipple, nipple-key, screw-driver, picker, a few spare caps, and a small steel oil-bottle. The ammunition, which for convenience in loading, and the certainty of ensuring always precisely the same charge—without which it is hopeless to expect accuracy of shooting—should be made up into cartridges; it may then be conveniently carried in the Spanish fashion, in a belt fitted with appropriate receptacles for holding each cartridge separately; the belt being made to pass round the body, and to be retained in its place by a buckle.

ARMS.

The merits of the various kinds of rifles now before the public, will be found fully discussed in my treatise on "The Rifle, and how to use it."* Having during the last twenty years tried every description of small-arm, and in so doing, having fired between sixty and seventy thousand rounds, my experience on the subject may naturally be presumed to exceed that of most people. From the frequency of the inquiry put to me—"What kind of rifle do you recommend?"—it is probable enough, that the form and description of many thousand stand of arms depend on the expression of my opinion on the subject, and I would therefore strongly urge attention to the following rules:—

1. The bend and length of the stock to be suited to the length of the neck and arm of the owner.

2. The bore of the rifle to be ·577-inch,† so as to

* A fourth edition of this work, revised and enlarged, has just been published by Messrs. Routledge, 2, Farringdon-street.

† The Government have it in contemplation to reduce materially the bore of all fire-arms; but this very desirable change will probably not be carried out for some years to come.

adapt it for carrying the regulation ammunition: this is indispensable.

3. The best form of breech-loader should be selected, in preference to muzzle-loaders; for this reason, that a body of men armed with the former, are at once a match, so far as rapidity of loading is concerned, for five times their number, if provided only with the old, clumsy muzzle-loaders.

The increased facility of loading, renders the breech-loader as superior to its rival, as the percussion rifle is in other respects to the antiquated fire-lock; there can be little doubt indeed but that in the course of a few years the use of muzzle-loaders will be entirely superseded in every European army.

For clubs, where rapid firing is of no consequence, and where the rifle need only be regarded as an instrument for educating the eye, and enabling it to act in concert with the hand, which is in fact the sole object of target-shooting, any tolerably-made rifle will, of course, answer the purpose, but its power and accuracy should be properly tested by some qualified person.

I have one, made for me last year by Daw, of Threadneedle-street (who manufactured nearly the whole of the many experimental arms used by the late General Jacob), and in justice to the maker, I must say, that I have never tried any gun that pleased me better.

On the other side is a diagram, showing the practice made with it at 500 yards, the wind blowing strongly across the line of fire. If any gunmaker can turn me out a weapon to surpass this, I shall be happy in my next edition to give publicity to the fact. In the meantime, I am perfectly content with this rifle. Mr. Daw informs me that he can manufacture a plain serviceable weapon, similar in all

Target, 6ft. by 2ft. ; 8in. bull's-eye.

Wind.

DATE.
16 October, 1859.

NAME.
Hans Busk.

DISTANCE.
500 yards.

WEATHER.
Clear, Windy.

NO. OF ROUNDS FIRED.
36.

NO. OF HITS.
32.

DESCRIPTION OF RIFLE.
Jacob's, made by G. H. DAW, Threadneedle-street, London.

DESCRIPTION OF AMMUNITION.
Conical bullet, flat base. (32.)

OBSERVATIONS.
30 rounds, fired from rest.
6 from the shoulder.

essentials to this, for 4l. 10s.; or he can supply at the same price, if preferred, an Enfield rifle with bayonet complete, far better finished of course, than the common regulation arm.

An excellent double rifle, by the same maker, with sights adjusted up to 1500 or 1800 yards, can be made for 10l.

Annexed is a representation of such a gun.

The Enfield rifle is well enough in the rough hands of the Militia and the Line, but to the members of Rifle Clubs, and more especially to Volunteer Corps, a more delicate, and at the same time far more formidable, arm might be advantageously entrusted. Take, for instance, Colt's revolving rifle, which at once gives every man provided with it, the efficiency of five men armed with the bungling and clumsy muzzle-loader. It is light, compact, well-balanced; the workmanship plain, but excellent; the finish all that can be required. In range and accuracy of fire, it far exceeds any other description of re-

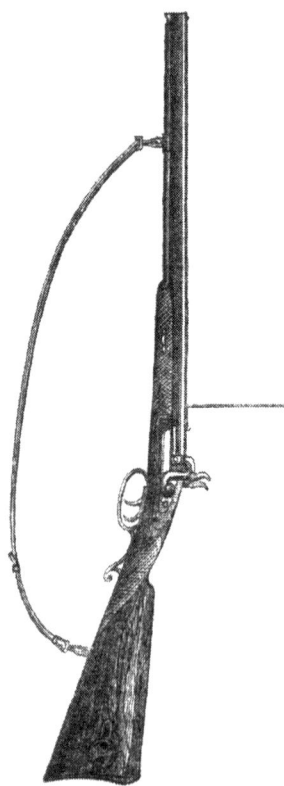

DAW'S DOUBLE RIFLE.

ORGANIZATION OF VOLUNTEER CORPS. 21

volver; though excessive range now so much talked of and vaunted, is in fact, a quality greatly overestimated in small arms.

But the particular kind of revolver above alluded to, has this great advantage over every other that has been sent to me for trial, that it is perfectly safe, and free from the many accidents to which most of the very questionable imitations of it are liable.

I have shown elsewhere, that Mr. Prince's rifle is undoubtedly at present the form of arm upon the breech-loading principle, best suited either for a foot or horse soldier. As the muzzle-loading system will soon become antiquated, it should be abandoned without delay, if we wish to keep pace with the age; rifles, either revolvers or breech-loaders, should be given, from time to time, as prizes to those who distinguish themselves as picked shots, uniformity of gauge being, of course, observed throughout every branch of the service.

With Prince's carbine, weighing $5\frac{1}{2}$ lbs. (which may easily be used by cavalry with one hand), a man may deliver his fire with ease and precision at the rate of eight or ten rounds a minute. He can load and "cap" with the greatest facility, in any position, and without that strain of the arm which so greatly disconcerts the subsequent aim when an ordinary muzzle-loader is used.

A handful of sharpshooters thus armed could, if judiciously posted, pick off every man at one of Armstrong's guns before they could fire two rounds, and with scarcely a chance of being touched themselves. The adoption or rejection of the breech-loading principle will test the value of the opinion of the Smallarms Committee, to whose judgment the question has been submitted. I am not, however, very sanguine in anticipating a beneficial issue from any military committees: their recommendations not unfrequently end in results which would be ludicrous, were they not so extremely costly; while they are besides too often utterly valueless. To take one instance from among many.

Not very long since, the inventors of various new and improved forms of rifled fire-arms went to considerable expense in preparing specimens of their guns for trial. An inquiry, extending over some months, ended in the selection of one of these, the least entitled to commendation—Sharp's American rifle. A rifle which every one, at all conversant with these implements, knew full well had been decidedly condemned by the American military authorities themselves years ago! Indeed, during a long course of experiments in America, it was found that independently of the very serious evils attendant on its mechanism, this carbine would not bear comparison with one of Colt's 12-inch pistols. At 200 yards, out of twenty shots fired from a rest, several missed an eight-foot target. At 300 yards, twenty per cent. failed to strike the same object. Yet here, a military committee not only selected this carbine as the special weapon for our dragoons, but ordered *seven thousand* of them at once, at a cost of about six guineas each! A very slight trial in actual service in the Crimea, where the men literally threw them away, proved the unfitness of the weapon for its intended purpose; and probably it is not too much to say, that none of these arms will ever be issued again. So much for a whim, which has entailed upon the country an absolute waste of between forty and fifty thousand pounds!

This expensive lesson at any rate proves that we must not too readily adopt the suggestions of military men even on such points. Nor are other instances wanting to prove the incompetence of committees similarly constituted. In conformity with the recommendation of a synod of the kind, it was determined some years ago, by our sapient authorities, to arm certain picked regiments with the two-grooved rifle. I had, long before, at considerable expense, tested that form of fire-lock in every way, and so well satisfied was I of its miserable inefficiency, that I took an opportunity of pointing out its manifest defects to some influential, and, as they were reputed, scientific military men, who appeared to acquiesce in the soundness of the objections. Not much to my surprise, however, I subsequently found that a large quantity of these useless arms had been constructed and issued, at an expense, as I was informed, of about 60,000*l.*, but within a few months they were all called in and condemned.

To sum up in a few words my recommendations to Rifle Corps as well as to Rifle Clubs, I would strongly advise, in all cases where a supply of rifles is to be bought, the adoption of one uniform description, if not of arm, at least of bore. In order to prevent confusion, let that, as already suggested, be ·577-inch. The length of barrel is not of quite so much moment, though it will be prudent to adhere in this respect to the regulation as well. With rifles upon Jacob's principle, a 24-inch barrel will give as great a range as the long Enfield; but where men are likely to have to fire in double files, the rear rank men, unless cool and well-disciplined, would be very apt to injure those before them. Besides, the regulation sword-bayonet cannot well be used with so short a barrel as that recommended by the late General Jacob.

Before dismissing the subject of Rifle Clubs, and in order to show what may be accomplished in a short time by men of the right sort, actuated by a proper spirit, and possessing tact and energy, I cannot do better than give here a brief sketch of the establishment of the CAMBRIDGE RIFLE CLUB.

On the 6th of May, shortly after the publication in the *Times* of some observations of mine on Rifle Volunteers, a public meeting was convened, chiefly on the instigation of the Rev. W. Emery, of Corpus College, at the Guildhall, at which the following resolutions were passed:—

1. That this meeting is of opinion, that in the present aspect of public affairs it is desirable to form Rifle Clubs, and that in accordance with this view it determines to form a Rifle Club for the University and Town of Cambridge.

2. That steps be taken to obtain the sanction and assistance of the Government to enrol such members of the Club as desire it into a Cambridge Rifle Corps, to consist of an University and a Town division.

3. That a donation list be at once opened to defray the preliminary expenses of the Club, and that a donation of 5*l*. 5*s*. constitute life-membership. That all other persons who desire to become members of the Club, whether practising or non-practising, pay an entrance fee of 10*s*. 6*d*., and an annual subscription of 1*l*. 1*s*.

In a few days the following rules were agreed to; and as they have been very judiciously settled, I give them here in full, as with slight modifications they may be adopted for other Corps:—

Rules agreed to at a Meeting of the Committee of the Cambridge Rifle Club, held in Sidney College, on Monday, May 16, 1859.

1. The affairs of the Club shall be managed by a Committee, who shall select a President, Vice-President, two Secretaries, and a Treasurer, annually from their own number.

2. The Committee shall consist of twenty members, to be elected annually, ten from the University and ten from the Town. The Committee, on resigning office, shall return eight from their number to serve for the ensuing year. The remaining twelve shall be elected by the members at large, on the first Tuesday in the month of May.

3. A meeting of the Committee shall be held for the transaction of business on the first Tuesday in every month. The Secretary shall call extraordinary meetings on the requisition of five members of the Committee, and due notice of such meetings shall be given by the Secretary to each member of the Committee. In case of an equality of votes, the chairman of the meeting shall have a casting vote.

4. (1) An instructor in musketry, (2) a bugler and armourer, (3) as also a marker, shall be appointed as occasion may require, chosen by the Committee. These officers shall always be in attendance during the hours of practice.

5. Each gentleman, upon entering the Club, shall immediately pay into the hands of the Treasurer the sum of 10s. 6d. as entrance fee, and one guinea annual subscription; or in lieu of both, a sum of not less than 5l. 5s., which shall entitle him to life-membership without further payment. All subscriptions shall be paid in advance, and shall become due for the ensuing year on the last day of May. The first year shall end on the last day of May, 1860.

6. No person shall be considered as continuing a member, whose subscription is more than two months in arrear, nor be allowed to shoot for prizes until his subscription and all other sums due to the Club be paid up in full.

7. All members of the Club who desire to receive regular instruction in musketry shall enter their names in a register to be kept for the purpose, and

by so doing they shall engage to submit to the following regulations:—

(a) There shall be three classes of such members.
(b) All such members shall begin their practice in the third class.
(c) The third class shall commence their practice at 50 yards, and no member of it shall proceed to the second class until he can put ten bullets out of twelve into a 12-inch bull's-eye at 50 yards.
(d) The members of the second class shall commence practice at 100 yards, proceeding to 150, and thence to 200. No member of the second class shall be promoted to the first, until he can put eight bullets out of twelve into a 12-inch bull's-eye at 200 yards. There shall be other stations at distances of 50 yards each, and in no class shall any member proceed to a new station till he can put eight bullets out of twelve into a 12-inch bull's-eye at the station he leaves.
(e) Any member of the second class who makes a hit off the iron target (6 ft. square) at or under a range of 100 yards shall be fined sixpence. The same penalty shall be inflicted on any member of the first class who hits off the iron target at or under a range of 200 yards.
(f) There shall be a register of attendance and performances of the members of the classes.

8. Members of the Club who do not wish to belong to the classes of instruction shall be at liberty to practise, at the times not occupied by the classes. A list of fines applicable to such members shall be drawn up by the Committee.

9. The maximum distance at which any member

ORGANIZATION OF VOLUNTEER CORPS.

shall be allowed to practise shall be fixed by the instructor in musketry.

10. If a rifle go off by accident in the hand of any member, that member shall be fined half-a-crown in all cases, and no excuse whatever shall be admitted.

11. If during practice, any one fire from a point not marked as a practising station, he shall be fined half-a-crown.

12. If during practice, any one stand in advance of the line marked as the practising station, he shall be fined half-a-crown.

13. The above fines shall be devoted to a fund for Prizes.

14. There shall be a challenge gold rifle, to be shot for at a half-yearly match, and held during the ensuing half-year by the member who proves himself to be the best shot at such match. Such prize rifle shall be presented to any member who may win it four times in succession.

15. There shall be prizes, to be shot for once a month, for each of the instruction classes.

16. The times of practice shall be regulated by the Committee, according to the season of the year.

17. The last Tuesday in each month shall be considered the regular shooting-day for medals and prizes. The conditions of entry and other necessary regulations shall be determined from time to time by the Committee.

18. The members of the Club shall not wear any distinctive uniform.

By Thursday, the 19th May, nearly 1000*l.* having been subscribed, and several hundred members being enrolled, the first practice took place on a temporary ground selected for the purpose. Thus, in ten days all the preliminaries had been settled, and a model club was already in working order.

SECTION II.

ON VOLUNTEER CORPS.

THE peculiar functions of Light Infantry in actual service are not only varied in their character, but of the utmost importance to the safety of the army to which they are attached. When troops are upon the march, it is the peculiar province of the Light Infantry to reconnoitre the country through which it has to advance, to clear the way for the heavy columns, to protect them from being harassed while retreating, to cover the movements of the line, keep a vigilant eye on the enemy, and test the character of the roads through which a march has to be conducted. Upon the activity and efficiency of the Light Troops it is, that a general has frequently to depend for the information requisite to enable him to regulate the movements of the main body.

The duties of Light Infantry are especially those for which Volunteers should qualify themselves. Too much attention therefore cannot be paid to the very simple drill detailed in the following pages. Considerable bodies of such troops, combined either with the solid column in the field, or with the extended line, would inspire mutual confidence. The regulars would be satisfied, that all the harassing duties of skirmishing, &c., would be well performed by Volunteers, while they, in their turn, would know that they had in their rear, a body of troops *en masse*, behind whom, when closely pressed they could retire. One important duty would be, to hang upon the skirts of the enemy, compelling him to be constantly on the watch, and continually harassing him with the apprehension of an attack. If Light Infantry discharged their duties efficiently, no detachment or reinforcement could be dispatched, no movement of consequence effected on the side of the enemy—scarcely, indeed, could a return be transmitted, or even a messenger sent off, without information being conveyed in a variety of ways to the head quarters of the army from which these light troops were detached. Rapidly dispersed in all directions, their effects are sensibly felt at the same time at distant points, and they tend to convey the impression to the mind of the hostile commander, that he has perpetually fresh obstacles to encounter.

Rifle Volunteers would upon almost all occasions have the advantage of choosing the time and point for their attack, which could only be resisted at a disadvantage, not only at the part where the enemy was weakest, but where the ground was most difficult to hold. If ably led, they would frequently be enabled to penetrate through his flanks, to the very rear of the enemy, stealing their way through his outposts, and, after inflicting severe injury, making good their retreat.

In the event of a doubtful conflict, instead of permitting the enemy to rally or retreat at leisure, a corps of active riflemen might frequently turn the scale of victory, as they have often done ere now. Should the enemy be retreating hurriedly, they close in upon his heels, they follow him up sedulously, allowing him no rest; should he, galled by his assailants, turn and form, they, on the other hand, retire, disperse, and disappear.

Great as formerly was the efficiency and importance of light troops, the adoption of the rifle has immeasurably increased their power. They should for the most part be able marksmen—men who could appreciate and render available the vast range of modern small arms. Where troops act in masses, it is not of so much consequence that they should have a very perfect arm. Soon after the commencement of any engagement, the intervening space between the contending forces is so obscured with smoke, that the combatants are utterly unable to discern any object at a few yards distance. Light Infantry, on the other hand, dotted here and there, at a distance from each other, selecting suitable cover, and able to use the stem of a tree, a rock, or a wall as a rest, ought to make certain of hitting any man they can discern, at a distance not exceeding 500 yards, especially when provided with a weapon capable of killing at four times that range.

In former wars, skill in shooting was little heeded, and even the most rudimentary knowledge of the principles on which all correct firing depends, were never so much as explained to the soldier, probably from not being understood by his superiors.

An officer of undoubted veracity once stated, that a party of Light Infantry under his command in Egypt, to his certain knowledge, fired away 60,000 rounds in an attack in which he was firmly convinced not more than four or five of the enemy were killed or wounded!

But all this is now changed. Such is now the precision of modern arms that in all future warfare there will necessarily be a much greater destruction of human life, in proportion to the ammunition expended, than was ever known in former times.

In 1808, under the apprehension of an anticipated

invasion, a voluntary armament of the British people took place, displaying a spirit of patriotism equal to that of the greatest nations at any epoch of the world's history. After an interval of fifty years, events, the purport and ultimate tendency of which it is impossible to mistake or over-estimate, have brought about a similar state of things; proving, happily, that the indomitable character of the nation has undergone no diminution in the interval.

Those among our military men whose experience in actual warfare is small, and knowledge of history still less, are too apt to speak sneeringly of volunteer levies, and to underrate their importance. The records of the past, however, show that on many occasions upon which Volunteers have been opposed to paid forces, the latter have not been able to stand against them. To enumerate a few instances, some of which I have already cited elsewhere, the reader may here be reminded that the Cevennois, in that short but desperate struggle carried on for two years against all the power, and skill, and generalship of France, never allowed four experienced marshals who were successively employed against them, to snatch so much as a single victory. These brave Volunteers did not at any one time number more than 3000, and yet in that disastrous inter-necine war, more than 50,000 picked veterans perished.

At Saratoga, in 1777, an army composed to a great extent of Rifle Volunteers compelled General Burgoyne to surrender with all his forces.

Another British Army, in 1781, under Lord Cornwallis, underwent a similar humiliation. Who has not dwelt with admiration upon the deeds of valour achieved by the German Student Volunteers in the wars that immediately preceded the downfall of the first Napoleon? The determined Circassians, badly armed and indifferently equipped as they are, have manfully held their own against all the might of one of the greatest military Powers in the world.

Major-General Sir W. F. Williams, before his departure for Canada, attended a public meeting at Liverpool, held for the purpose of organizing a Volunteer Corps there. The high opinion entertained by that gallant officer of the value of these forces will be seen on perusal of the following extract from his speech on the occasion. He combated the idea prevalent in some quarters, that in raising corps of this description "they would be placing weapons in the hands of those whom they might have cause to fear. God forbid that he should entertain such a feeling. He was sure that every rifle placed in the hands of Englishmen would be only used in the defence of their Queen and country. (Cheers.) With regard to the river, there were hundreds, he might almost say thousands, of boats upon its surface, every one of which was capable of being armed in a most efficient manner; and few would imagine what a swarm of hornets might thus be sent round the heads of an enemy. He would warn them against supposing that Volunteer forces could alone suffice to stop an invading army, but they would be most useful as auxiliaries to a regular army. The surface of England, intersected as it was by numberless enclosures, was wonderfully adapted for the action of that particular description of Corps which the country are now about forming in such numbers. With adequate numbers of regular troops acting in conjunction with a large Volunteer Rifle Corps, they might snap their fingers at the united world."

Properly organized and trained Rifle Volunteers would be found, too, in every way far better adapted than other troops, to quell popular tumults; while their

VOLUNTEER REGIMENTS. 33

great use would be shown in any sudden emergency, more especially on the landing of a hostile force upon our coasts. The last time these Corps were called out, no inconsiderable pains were taken and expense incurred to render them efficient by means of thorough drilling.

By attention, however, to the system now laid down, and detailed in these pages, instead of a long and irksome course of training, a few hours daily, for three or four weeks, will suffice to qualify every man for the irregular forces now forming over the country. The main instructions extend to marching, facing, and wheeling alternately to the right and left, loading and firing quickly, but without hurry, breaking into independent files, or extending from the right, left, and centre, re-forming as rapidly. A few simple bugle calls must also be learned, together with the mode of forming divisions and subdivisions, and doubling rapidly from the right or left without confusion, in passing an obstacle. These are the principal manœuvres, and they should be thoroughly mastered before proceeding to the others subsequently detailed at length.

Our Volunteers, too, might advantageously practise an evolution which by its suddenness would on various occasions strike terror into an enemy, who, if it were rapidly performed, could scarcely oppose it with effect. For this purpose, while skirmishing as light troops, they might, if a favourable opportunity occurred, and they felt themselves sufficiently strong, form in column upon a preconcerted signal, and charge the enemy's line.

After some little practice, such Volunteers would be of great service in concealing the movements of large detachments of regulars. As a general rule, the movements of all light troops should be on the centre, the shorter the distance they have to traverse the better, where rapidity is essential, and *" Celer et audax"* must be their motto. It is also far easier to form and dress on the centre than on either of the flanks, which of course are at double the distance.

A thorough knowledge of the country is of the greatest value to all troops; to none more so than to riflemen. Now, our local Volunteers, living almost on the very spot, where if their services were needed they would have to fight, would be intimately acquainted with every road, path, wood, copse, and stream for miles around them. An army utterly ignorant of the locality where they may be encamped, could not venture from their entrenchments surrounded by such a foe without the greatest risk of being cut off by invisible riflemen, who from every convenient place of concealment, could keep up a destructive fire with but little hazard to themselves.

In this kind of guerilla warfare, good marksmen are doubly formidable; none but skilled shots can be opposed to them, and he who is most proficient in the use of the rifle has the best chance of carrying the day.

Nothing besides gives such thorough confidence as a perfect acquaintance with all the intricacies of a country, its bye-ways, hedge-rows, and defiles. In one so enclosed in all directions as our own, should an enemy land, there would be little opportunity for any conflict like a general engagement unless they should succeed in making their way unopposed to such an open expanse as is to be found in the vicinity of Aldershot or Salisbury. The very character of our southern counties, renders fortifications superfluous; for, be the invading army never so numerous, it would have to advance along one continued line of banks, breastworks, hedges, copses, gullies, &c., all of which would be lined with cool and determined riflemen.

Supposing some of our riflemen to be driven from their first line of breastworks, they would have only to fall back, firing as they retreat, posting at the same time a few good marksmen under cover, in such a position as would enable them to flank their pursuers, and as soon as they have reached another bank or breastwork, the enemy would have all their work to begin again.

When riflemen are dispersed in different directions in small parties, they can venture more boldly forward, and encounter greater risks, being certain then of speedy succour, while the enemy, should he attempt to charge them, runs the risk of being entrapped in an ambuscade.

Were all our Volunteers indeed trained as light troops, to do patrol duty, and had *we plenty of them*, they would enable much larger numbers of the regular infantry to be brought into the field. No inconsiderable portion of every army is required for these laborious occupations, generally rendered all the more harassing from the want of knowledge of the shortest roads and most important posts. To troops, on the other hand, possessing the intelligence and spirit of Volunteers, such services would be comparatively light. These facts are so well known to all military men of any real experience, that they are always found, to a man, warmly advocating the permanent adoption as an institution, of an ample number of Volunteer regiments. It is only the timid and effete, the class who glory in pipe-clay and stocks, who clung fondly to "Brown Bess," and even now prefer that antiquated tinder-box to the percussion rifle; or else the very young and *dilettante* soldier, who has the rudimentary principles of his profession to acquire, whose voice is ever heard, in opposition to the advantage of Volunteer forces.

The present Government has exercised a most wise discretion in not dallying, as their predecessors had frequently done, with the probabilities of danger from foreign aggression, which, obvious enough years ago, have only just been discerned and admitted. The circular issued a few days since from the War Office is decidedly one step in the right direction; still it will be of little value if not followed by other measures.

General Peel, in calling attention to the provisions of the Act of 44 Geo. III., cap. 54,* hesitated to take upon himself the responsibility of pledging the country to any expenditure on account of Volunteers at a time when Parliament was not sitting. Still, if the country is to have the great benefit, which every man of common sense must see will accrue, from having a large body of skilled marksmen and well-trained Light Infantry disposable on all occasions, at a few days or hours notice; surely that very large class who, from age, infirmities, or other causes, cannot be personally enrolled, ought to be ready and willing to contribute to some extent, to the unavoidable expenses attendant upon the formation and maintenance of a means of defence so essential to the safety of the empire.

At present the country annually disburses 88,000*l.* towards the expenses of Yeomanry Corps. Without proposing to discuss here, whether or not an equivalent be received for that large expenditure, it may be confidently affirmed that a similar sum, properly laid

* An abstract of this Act, together with the subsequent enactments relating to volunteers, is given at the end of this treatise.

out, would go far towards covering the cost of providing arms, practice-ammunition, and accoutrements for those who may be unable to bear the outlay, and are yet willing enough to serve in the ranks of Rifle Corps.

It is incumbent upon the Legislature, as soon as it re-assembles, to take this very important matter into their consideration, and to re-model at once the old statute of 1804, adapting it to the exigencies of the present occasion, that is, if there be any real desire on the part of our authorities, to turn to account the loyalty and good feeling so largely evinced of late in every part of the kingdom.

What the Government ought, among other things, to do, is, to confer some slight advantages upon all who volunteer in the service of their country, instead of the ridiculous boon now conferred by the Volunteer Act—namely, an exemption from the duty on hair powder! Surely some better encouragement than this could be devised; for example, an immunity from serving on juries—from the duty on game certificates—from the tax on armorial bearings, or some other trifling privilege of that description, which would far more than make up to the community for the insignificant diminution to the revenue thereby occasioned.

The pay of the Adjutants, serjeant-majors, armourers, and markers of Volunteer Corps ought, unquestionably, to be defrayed by Government. A certain number of stand of regulation arms should be distributed to each regiment, and most decidedly the members of a corps should not be compelled, as they now are, to purchase their own ammunition.

Whether or not those in the humbler classes of society, who would be unable to give their time gratuitously, should be paid, is another question; it is, however, very much to be doubted if any merely

38 RIFLE VOLUNTEERS.

A STATEMENT *showing the total Strength of the Land Forces available for the Service of Her Majesty, for the Years 1859-60—exclusive of Militia and Volunteers.*

DESCRIPTION OF FORCES.	Officers.	Non-commissioned officers.	Rank and file.	1859-60.	
				All ranks.	Horses.
CAVALRY.					
Royal Horse Artillery	49	96	1,371	1,516	1,200
Life Guards and Horse Guards	99	162	1,053	1,314	825
Cavalry of the Line	476	687	8,078	9,241	5,993
INFANTRY.					
Royal Artillery	657	1,296	15,224	17,177	2,800
Royal Engineers	368	293	3,008	3,669	120
Military Train	107	156	1,438	1,701	1,000
Foot Guards	261	439	5,600	6,300	...
Infantry of the Line	2,970	5,295	62,200	70,465	...
Medical Staff Corps	2	70	928	1,000	...
West India Regiments	180	239	3,000	3,419	...
Colonial Corps	249	410	5,140	5,799	900
Total, Regimental Estabs. . .	5,418	9,143	107,040	121,601	12,838
STAFF, &C.					
General Staff (exclusive of Officers on Regimental Full Pay)	120	...	...	120	...
Commissariat Staff	206	...	...	206	...
Medical Staff	338	...	...	338	...
Staff of Depôt Battalions, &c. .	108	98	...	206	...
Staff of Recruiting Districts . .	35	68	...	103	...
Commissioned Chaplains . . .	81	...	...	81	...
Total, Home Forces	6,306	9,309	107,040	122,655	12,838
Add: Her Majesty's British Forces in the East Indies—viz.,					
Royal Horse Artillery . .	29	52	748	829	680
Cavalry of the Line . .	410	649	7,128	8,217	7,315
Royal Artillery	261	489	7,128	7,878	2,568
Royal Engineers	19	32	458	509	...
Military Train	18	33	236	287	250
Infantry of the Line . . .	2,776	5,138	66,200	74,114	...
Medical Staff Corps . . .	...	6	57	63	...
Total	3,513	6,399	81,955	91,897	10,813
Depôts of Regiments in India stationed in Great Britain .	430	1,067	13,508	15,005	418
Total Force on the East Indian Establishment, including Depôts in Great Britain .	3,973	7,466	95,463	106,902	11,231
Total of her Majesty's Forces .	10,279	16,775	202,503	229,557	24,069

VOLUNTEER REGIMENTS. 39

voluntary and unpaid system will produce either permanence or numbers, and, in the face of the great Armies of Europe, it is only by very large numbers that Volunteer Corps would be either effective or useful.

The Armies of the different States of Europe at this moment, upon a tolerably close approximate calculation, exceed 3,400,000. The available military power of France alone exhibiting 536,000 men.

On the opposite page is given, in a tabular form, the exact details of the forces at her Majesty's disposal during the current year. The reflection naturally arising from an inspection of the comparative insignificance of England as a military power, will, of course, be balanced by a consideration of the magnitude* and extent of the very judicious naval preparations now making; but one would imagine that the most stolid and infatuated amongst us, could hardly fail to acknowledge the expediency, if not the absolute necessity, of adding largely to the number of her Majesty's forces by very considerable Volunteer levies. A few clubs here and there—a few Corps, some 300 or 400 strong, are very well in their way, as the *nuclei* of greater things; but Parliament must forthwith lend a hand, and an earnest hand, too, if any real good is to be achieved. We ought, if matters are not sadly bungled or mismanaged, soon to have 100,000 men learning their drill, and acquiring proficiency with the rifle, and before the end of the year 100,000 more should be coming forward to swell the ranks. I have lately been favoured by a gentleman of great and well-known ability, a member of the Legislature, with an original scheme, which will probably be brought forward, ere long, in

* See "The Navies of the World." Routledge. Published May, 1859.

the form of a Bill. From the care with which it has been prepared, it is entitled to every consideration.

The following sketch of his proposal for raising Volunteer Corps is based upon principles which do not seem to have been regarded in other schemes having the same object in view. In his opinion:—

1. The arrangement should be permanent.
2. The effective result should be cumulative from year to year.
3. The remuneration to be given should depend upon the engagement being completely performed.

England should be divided in the first instance into about 400 districts, for the purpose of raising one regiment from the inhabitants of each district.

Each district should be formed with reference to the density and character of the population, so as to afford the greatest facility for the Volunteers being assembled for duty.

The strength of each regiment should also be determined by the density of the population, and their aptness for service. The minimum being five companies of about eighty men each, and the maximum ten companies of about 100 men each.

The thinly populated districts should be divided into sub-districts for each company.

For each regiment there should be a permanent staff appointed by the Crown, composed of one Adjutant, one serjeant, one bugler, and one drill serjeant for each company.

Some of the staff might be selected from the pensioned non-commissioned officers of the line as a reward for meritorious service, and they should receive the pay of their rank.

The Crown should provide a house in the centre

of each district, to be appropriated for the purposes of a residence for the Adjutant, an office for the business, and a store for the arms, uniforms, &c., of the regiment.

In sub-districts a house should be provided for the same purposes for each company, but the drill-serjeant of the company should reside in it.

This staff might prove extremely useful in enlisting recruits for the regular army, without at all interfering with its primary duties in the Volunteer Corps.

The Crown should be empowered to appropriate any open space contiguous to a town or in the country suitable for drilling, and for ball practice, for each regiment, subject to such restrictions as might be found necessary.

The officers might be nominated by the Secretary-at-War, and they should be required to qualify themselves within twelve months for the discharge of their duties at their own cost.

On obtaining a certificate of competency from the Adjutants of their respective regiments, they should receive commissions from the Crown.

For each regiment there should be one Colonel, one Lieutenant-Colonel, and one Captain, one Lieutenant, and one second-Lieutenant for each company.

After the first formation of the corps, all officers should enter as second-Lieutenants, and be promoted by seniority, provided they obtained certificates of competency at each promotion from their Adjutant.

Each officer should be required to attend a certain number of parades yearly on pain of removal.

The officers should provide themselves at their own expense with uniforms, arms, and accoutrements.

Facilities should be afforded for enabling officers to

learn their duty by their being allowed to attach themselves to regiments of the line for instruction, as also by the establishment of training corps at the two universities.

Officers should be allowed to resign at pleasure, and on changing their residence to another district, should be transferable to the regiment of that district as supernumeraries to fill the first vacancy of their rank below that of lieutenant-colonel.

The non-commissioned officers should be selected from the privates by merit, and in all other respects should be on the same footing as the privates.

The privates should, on the first formation, be selected from the population of the district between the ages of 16 and 30, to such an extent as would form the nucleus of the regiment without interfering with the industrial requirements of the district.

In the first year the privates should receive sufficient training to acquire a general knowledge of their duty. For this purpose it would be necessary to make special arrangements suitable to the circumstances and occupation of the people in each district.

In each year afterwards, a certain number of youths of the age of 16 should be enrolled in each district, so as gradually to increase the strength of the regiment to its full number, after which the district should be divided and a new regiment formed.

The recruits so enrolled, should be embodied and trained for about four months in their first year.

The regiment, after the first year, should assemble for an hour or more daily for a certain number of days in each year, for the purpose of training and for ball practice.

Each non-commissioned officer and private should be required to attend an hour daily for about thirty days in the year for training, and four days for ball practice, unless excused by his commanding officer on the ground of ill-health.

The arrangements for training should be made with a due regard to the circumstances and occupations of the privates of each regiment, so as to interfere as little as possible with their ordinary avocations.

In thinly populated districts the men should only assemble in companies for training and ball practice, except for three days in the year, when the whole regiment should be mustered.

The Crown should provide the arms, ammunition, and uniforms as follows:—An old rifle or musket, and sword-bayonet to each man for drill; one well-finished regulation rifle to every ten men for ball-practice; a strong serviceable blouse and leathern belt, with ammunition pouch and sword belt; a forage cap, with band (on which the name or number of the regiment should be marked), for each man.

The arms and uniforms should be kept under the charge of the Adjutant, and in sub-districts of the drill-serjeant.

The Crown should also provide ammunition for ball-practice.

The privates should be engaged to serve until they reached 50 years of age, but they should be entitled to their discharge at any time on forfeiting all remuneration, except in case of impending war.

Should any private remove to another district, he should be entitled to be transferred to the regiment of that district.

If a private should enlist in the regular army, he should be entitled to his discharge, and should be allowed a certain period of service, in proportion to the time he had been enrolled as a volunteer.

In case of impending invasion, the regiments should be liable, at the call of the Legislature, to be embodied, and be required to do duty in any part of the United Kingdom, but should not be liable to serve beyond seas.

The privates, whilst being instructed in their duties during the first year of their enrolment, and upon entering into an engagement to serve for a specified period, should receive suitable pay.

On the regiment being embodied for service, the officers and men should receive the same pay and allowances, and be in other respects on the same footing as the embodied militia during the period of service.

On fulfilling all the conditions of the service, the non-commissioned officers and privates should at the age of 50 be entitled to pensions. The non-commissioned officers to 20*l.* a-year, and the privates to 16*l.* a-year.

In the case of temporary absence, or other imperfect performance of service, where an entire forfeiture of pension might be deemed unduly severe, the pension should be reduced in amount, according to the degree of default.

The pensions of the privates first enrolled, should be subject to a deduction of 10*s.* for each year of their age above twenty years.

Along the coast, batteries of artillery should be established in the same manner, instead of rifle corps.

A naval reserve might also be provided from the seafaring population by the application of the same principles. It would only be necessary that youths should serve for about two years in the Navy, and should occasionally be trained on board the nearest vessel of war.

VOLUNTEER REGIMENTS.

Seven years ago Mr. Henry Drummond, M.P., published a very useful pamphlet on the subject of Volunteers. It was alluded to recently in the *Times*, but as it has been long out of print, and as many of the observations it contains are peculiarly apposite at the present juncture, I have ventured to condense the most important into a compact form, here subjoined. It will be seen, from Mr. Drummond's own statement, that his experience entitles him to claim attention to his remarks.

"Few persons," he observes, "have seen more of Volunteer regiments than I have done. A long time ago, I served for two years as private, in the light infantry company of the Oxford Volunteers. Subsequently I was captain of a troop of yeomanry in Hampshire. I afterwards commanded a regiment of Infantry in London, about 800 strong, from which I changed to the command of a regiment of Riflemen, 300 strong, where I remained until we were disbanded at the general peace. Latterly, I took some trouble in the raising of the Surrey Yeomanry, of which I was appointed Lieutenant-Colonel.

"After this experience, I think I am entitled to give an opinion upon what such forces can do, and upon what they cannot do; and whilst I strongly urge you to form a corps of riflemen, I must frankly tell you that no such troops will ever be able to supply the place of regular regiments, nor to face columns of the enemy.

"To show the way in which they may be useful, it is necessary to bear in mind the state in which the country will be, whenever an invasion shall take place; but there is a preliminary question to be settled in the minds of many persons, which is, Will the French Emperor really invade us or not?

"I think he will, for the following reasons:—1st.

It is his interest to do so. He has a large army of desperadoes under his command, for whom he must find employment. These men have been used to war in Africa, where they have been trained in acts of cruelty and villany, such as never before disgraced a Christian army, save that of Raymond of Toulouse, urged on by priestly fanaticism.

"In the second place, the French pant for war with somebody, and most of all with us. I have mixed much with the lower orders in that country, and sought to find out their real opinion; and I invariably found them longing for the Rhine as their boundary, and, above all, to avenge the battle of Waterloo and the occupation of France by our troops. The tradesmen in Paris and some manufacturers are the only people who wish for peace.

"If an invasion of England do take place, one of two things must result; either England must be made a province of France, or we must put to death every one of the invaders. It is probable that the French will employ 300,000 men upon this expedition, of which 100,000 will land in different parts of the south of England, and at different times, whilst nearly an equal number will be destroyed by our ships, the rest remaining at home as a reserve. Of those who land, we may calculate that 25,000 will get to London; that they will set it on fire in many places, and be ultimately destroyed from behind barricades, from houses, and various places, under shelter of which, the male inhabitants may protect themselves, and kill the invaders in detail. For there is this disadvantage for an invading army in a hostile country, which is, that it can never move except in masses. A regiment of cavalry raised in Tuscany, marched three times into Spain, by Narbonne, 1200 strong; it never was once engaged in a general action, yet it returned three times a skeleton under 300, having been destroyed by the peasantry in the mountains by all sorts of cruel devices. Of the 100,000 French who will land here, it is not likely that a single man will ever return to France; but we must calculate that we shall lose at least 300,000 Englishmen.

"I do not believe that Buonaparte ever had any intention of invading England; but that the army assembled nominally with that destination was a feint. I was in France during the Hundred Days, and met frequently the principal persons on the staff of all nations. There was a story current then, which assumed to have Marshal Ney for its authority, that a few days before the order was given at Boulogne to break up the camp and to march into Germany, Buonaparte called Ney to him and said, 'This was your plan; here we are, now tell me how do you mean to execute it? How many men do you want?'

"*Ney.* 'I shall require 200,000 men.'

"*Buonaparte.* 'Well, what will you do?'

"*Ney.* 'I will embark them, and I shall probably lose half on the passage; but I shall land 100,000.'

"*Buonaparte.* 'Well?'

"*Ney.* 'Then I shall march directly to London, for there is nothing to stop me but the lines at Chatham, and they are trifling, and the English generals are all old women.'

"*Buonaparte.* 'Well?'

"*Ney.* 'Well, then, I have succeeded.'

"*Buonaparte.* '*Bête!* I grant that there are no defences—that the generals are old women, and that you will get to London; but you are judging of England by France. Whoever is master of Paris is master of France; but whoever is master of London is not master of England. You will have to fight for every inch of England as hardly as you will have fought for London; the country will make no terms with you: not a man of your 100,000 will ever return to France to tell me how you have got on, for the very women would rise and kill your stragglers and wounded with their bodkins.'

"The chief danger to us arises from our want of taking proper precaution. We, with insular ignorance and insolence, despise all Frenchmen; and think we can easily repel them. They do not despise us, but they hate us with an intensity which is increased the more they see of us in their own country, and the more they live in this. All generals and admirals, who alone are competent to give opinions upon the state of our defences, have given us warnings of their inefficiency. Let us see, then, what is the state in which the country must be, whenever an invasion takes place.

"I take it for granted that the safety of the Queen is the point from which everything else must radiate. Every Frenchman in this country, from the princes at Claremont downwards, should be shut up in prison. Every man in Great Britain capable of bearing arms must be enrolled and drilled; and no exemption permitted to any class except medical men, and then only on certain conditions.

"If the Government be wise, it will recall our troops from the Colonies, and strengthen Gibraltar and Malta. It will add, at least, 10,000 men to our regular army, and call out the Militia. The local Militia is more expensive than Volunteers; and I have seen some regiments of Volunteers better than any local Militia.

"The use of Volunteers will be, to hang on the flank and in the rear of the invaders, harass them incessantly, shoot all stragglers, and never suffer them to rest, particularly at night. In war, more men die from fatigue, and wet, and cold, than from bullets; and while the regular troops impede the advance of the enemy, the irregular riflemen and cavalry may destroy more than the regulars themselves. Wherever there is a large body of these irregular troops attending the regulars, the object of the General will be, at all hazards, to break the line of the enemy, whereby he will be thrown into confusion, and whilst he is in confusion, and before he has time to re-form his line, the irregular troops are as efficient as the regular, nothing being then required but personal strength and courage.

"Such being the use of riflemen, the next thing to consider is, the way in which they should be equipped. When men volunteer to join the regular army, they are always young men of activity, health, and strength, all nearly alike in these qualities. But when the whole male population, under sixty years of age, is brought out to be disciplined, the case is widely different. Those who prefer to serve as riflemen, should be exempt from service in the militia; but as the price of obtaining this privilege, they should furnish themselves with all things necessary for their service. It is, therefore, a great object that the expense of equipment shall be as small as possible. It is necessary that their dress should outwardly be uniform, in order that they may be known by their friends and distinguished from their foes—a point not always easy, amongst masses of irregular troops in the hurry and smoke of action.

"To fulfil these conditions, the best dress for riflemen is a grey round frock, descending half-way down the thighs, with a straight collar to button round the throat; a broad, brown leather belt round the waist, to contain the ammunition, &c.; and a straight sword to be fixed on and used as a bayonet; a pair of duck trousers,* and high shoes. The advantage of this frock is, that a chilly man will be enabled to put on any extra clothing underneath, without disturbing the uniformity of the outward appearance, whilst those who are oppressed by heat may wear as little as they please. Each man must be provided with a grey Mackintosh cape, descending to the knees, to keep him dry, and to serve as a blanket at night: this is to be rolled up when not required, and slung behind the shoulders, and cover a linen bag containing a pair of shoes, a pair of stockings, and a flannel shirt. Nothing else should be required, but every one should be allowed to carry as much more as he pleases. In bad weather, this bag would be hung round the neck on the back under the cape.

"The commanding officers of rifle corps, whenever they march from home upon real service, should form their companies and troops so as to have all the best men in one, and all the worst in another, and so forth. They would thus be able to detach a good contingent, though the whole regiment might not be fit to be used. General Sir Charles Napier published a valuable letter, which every one should read, in which he enumerates the things necessary for such troops to do. It is comparatively easy to get irre-

* It has been judiciously suggested, that in lieu of trousers, the garment known as "Knickerbockers" might be advantageously substituted. These are long loose breeches, usually worn without braces, and buckled or buttoned round the waist and knee. They obviate the drag in the leg, caused by walking up hill with wet muddy trousers, and are in fact as convenient and sensible a covering for the lower limbs as could be devised. With them may be worn a stout-ribbed worsted stocking and ankle boot; but a gaiter or legging of cloth, leather or unbleached canvas, which could be readily cleaned, might be added.

gular regiments to go through their field exercises, but the things chiefly to be practised are, firing at a target—running at least a hundred yards in a minute—marching from ten to twenty miles a-day—moving in double-quick time over ploughed ground, &c.; all movements being done by the sound of the bugle.

"Every rifleman must have a straight sword, thicker at the hilt than at the point, and tapering. It must be fixed on the rifle so as not to disturb the sight, by having the catch underneath the barrel. Whenever the regiment is sufficiently steady to be able to charge with the bayonet, this is the best weapon with which to meet a Frenchman. It requires a high state of discipline to do this; and if it could be attained by raw levies, pitchforks and blades of scythes on poles would be more efficacious than muskets for them. Moving in line, and particularly at double-quick time, is one of the most difficult things to attain.

"The large houses of gentlemen must be arranged as hospitals for the wounded. Coarse sacks, seven feet long, four wide, and two deep, filled with straw laid across are the best, sweetest, and most easy beds on which any man can lie. I say *filled*—filled as full as possible.

"The Lord-lieutenant should register the names and addresses of all the medical men in his county, and appoint some one at their head to direct their attendance at the temporary hospitals. The Lords-lieutenants must also direct some deputy-lieutenants to take account of all the horses, waggons, spades, and pickaxes which may be found in each parish; and also the number of men capable of using them, who are too old to serve as soldiers. There is no country in which it is more easy to construct impediments to the march of an invading army than England, by cutting ditches, raising embankments, throwing down trees, &c.; all of which are fatal to artillery; and it is artillery that decides the fate of battles and of empires.

"The women may be of great use in bringing provisions to the troops that are engaged, and in receiving into their houses the inhabitants who have abandoned their homes in the line of the enemy's advance, and which will probably have been burnt to the ground. They will, of course, volunteer their services as nurses; but the farther they keep from the country occupied by the enemy the better. The horrors of war are sufficiently terrible for all; but they accumulate with unspeakable aggravation upon women. Few who have shared in them but reflect upon them afterwards with disgust; and it is those military men in England who know what the reality is, who earnestly implore us to be prepared to repel the invaders, should they come. *The surest way not to be invaded is to show such a preparation as shall make the enemy see that such an attempt is hopeless;* and one part of our danger arises from the activity, secrecy, and falseness of our enemy. He has repeatedly asserted, that he was the last man in France capable of violating the constitution, during all which time he was plotting how to do it: he outwitted the penetration of the sharpest men in the Chambers; out-generalled Changarnier, Lamoricière, and all the best generals; without a whisper of his intentions having transpired.

"He will send his troops, not after a proclamation of war, but during a proclamation of peace; and they will have landed before we have had the smallest intimation of his intentions.

"I do not believe it possible for an invading army

to conquer this country if we resist in the manner indicated :—first, because the sea enables us to prevent the advance guard of the army from falling back upon its reserve, and also from being supported by it; and, secondly, because we can certainly, sooner or later, bring at least a million of armed men down upon them. But, I need scarcely add the truism, that everything depends upon taking adequate measures, upon being properly prepared, and upon being determined never to make any terms, but to conquer or die."

There is much sound sense in the above suggestions; they are particularly valuable at this moment, when England is gathering up her strength as she did some sixty years ago. Under the expectation of the possibility of invasion in 1794, in addition to a very large army of regulars, and a Navy manned by 85,000 seamen, the British Government subsidized 40,000 Germans, raised her Militia to 100,000 men, and called out a noble array of Volunteers. Between 1798 and 1804, when these forces were of the greatest account, they numbered 410,000, inclusive of 70,000 Irish.

On the 1st January, 1804, as appears from an official return, there were in England alone, 341,600 Volunteers, the population then being about one-half of its present amount.

SECTION III.

ORGANIZATION OF VOLUNTEER CORPS AS DISTINCT FROM CLUBS.

THE VICTORIA RIFLE REGIMENT is regarded as the model corps, *par excellence*, upon which all those now being raised over the country will be formed. In describing its constitution, therefore, every requisite information will be given, in order that it may serve as a guide for the organization of other corps.

The regiment in question was enrolled for home service, in 1853, under the "Volunteer Act" of 1804. Its present complement is 300 men, in four companies of 75 each, but these numbers will probably before long be much increased. The following is the establishment of officers:—A Lieutenant-Colonel, one Major, four Captains, four first Lieutenants, four second Lieutenants, an Adjutant, and a Surgeon. None of these officers receive any remuneration, with the exception of the Adjutant, whose duties being continuous and somewhat severe, he is provided with a house and an annual stipend, in addition to his army half-pay.

The sergeant-major is constantly in attendance, to give instruction in musketry &c.; and a resident armourer, who is provided with a house upon the ground, has charge of the arms. The Corps has also the paid services of a bugler and marker; both formerly in the Rifle Brigade. The arms used are the short Enfield rifle, with sword bayonet. The uniform is black, and has been pronounced by com-

RIFLEMEN SKIRMISHING.

petent judges the neatest and most appropriate worn by any regiment. It is represented in the preceding page. While alluding to this subject, it may be observed here that an iron-grey would be about the best hue for the uniform of a rifle corps in actual service. The shako and plume are worn only on parade; the forage-cap, shown in the frontispiece, being the covering ordinarily worn at drill. The total cost of the outfit for a private is as follows:—

	£	s.	d.
Tunic and trousers	4	10	0
Forage cap	0	14	0
Shako and plume	2	2	0
Cross belt, with cartouche-box	2	5	0
Enfield rifle (better finished than the common regulation arm), with sword bayonet and scabbard complete	4	10	0
The members of this corps being all by birth and position *armigeri*, and as such, entitled to wear a sword when they parade without rifle and bayonet, are provided with that side arm, although this, it will be seen, is not a requisite, the cost with belt being	4	10	0
Total	£18	11	0

Should he think proper, the recruit may provide himself with a regimental great coat or cloak; cost, five guineas.

If to the above amount we add one guinea entrance fee, and two guineas for the subscription for the first year, paid always in advance, we shall find that each member disburses in the first instance rather more than twenty guineas. It will, however, be obvious that a very efficient corps, composed of men of an inferior social class, clad in some such loose garment as that described by Mr. Drummond, in lieu of a regular uniform, and armed with the common Enfield rifle provided by the Government, either gratuitously or at the cost price, say about 3*l*. 8*s*. or 3*l*. 10*s*., may be equipped at a much more moderate rate; viz., for about 8*l*. each.

The regulations of the Corps are necessarily framed strictly in accordance with the Acts now in force, the members being, to all intents and purposes, soldiers; and upon their admission, taking the same oath that is administered to all officers and soldiers in her Majesty's service.

A considerable number of the gentlemen who have joined within the last twelvemonth, and are now coming in, are candidates for commissions in the Army and Militia, who are thus laudably taking advantage of an admirable opportunity for qualifying themselves for their future regimental duties, by acquiring a thorough knowledge of their drill in company with others occupying the same social position. They thus obviate the necessity of going through a similar ordeal at a future time, in not very pleasant propinquity with the rank and file they will have to command.

Were our Military Authorities only aware of the very great advantage thus afforded, not only to the young officers themselves, but to the Service at large, they would not hesitate to make the production of a certificate of proficiency from the Adjutant of the VICTORIA RIFLES a *sine quâ non* on the part of every applicant.

This corps more especially claims notice, because it has now been established for a series of years, is thoroughly organized, and is fortunate in possessing in its Adjutant as active, intelligent, and efficient an officer as any in her Majesty's service.*

Many months must necessarily elapse ere any other corps, formed upon the principle of the one in question, could get into working order; and even then, there can be no probability that any *metropolitan* corps at least, would rival the " VICTORIA" in the high character of its component elements. The rank and file of this well-trained regiment now comprises a very large proportion of Oxford and Cambridge men, members of both Houses of Parliament, officers who have served in the army, a considerable number of members of the Bar, and not a few county magistrates. It is obvious, therefore, that in London and its vicinity, all gentlemen of any position, desirous of becoming Volunteers, will naturally desire to enrol themselves in what must be considered the " crack" corps, in preference to connecting themselves with any new and doubtful association, compounded probably of heterogeneous materials, which, after incurring heavy expenses, may not after all be found to work harmoniously, and may soon have to be wound up.

For the reasons already stated, no less than from its having taken the initiative long ago, there is no question that the VICTORIA RIFLES will continue to hold the same relative position to any new metropolitan corps of the kind, that the Guards do to the regiments of the Line, and that from its ranks officers will continue to be eagerly sought to take commissions in other Corps. The present high character of the Corps is always likely to be maintained, as recruits are only admitted by the officers, after ballot, as vacancies occur. Every candidate must besides be proposed by one member, and seconded by another.

* Lieutenant Trew, late of the Rifle Brigade, resident at the head quarters of the regiment at Kilburn.

VOLUNTEER CORPS DISTINCT FROM CLUBS.

A rifle battalion similarly constituted in some respects to the Victoria Corps was founded at Exeter, under the sanction of her Majesty, on the 26th of March, 1852. It is stated that three or four full companies might be raised in that city among a respectable class of young men, who, if not enrolled in this corps, are liable at any time to be drawn for in the militia. "If the employers of clerks, shopmen, warehousemen, &c., would arrange so as to give their young men six or seven hours a week for drill, it cannot be doubted," says a gentleman connected with Exeter, "that a very numerous and valuable class of volunteers would enrol themselves." With reference to the expense, one of the officers (Captain Moore) stated that the uniform, accoutrements, &c., would cost 6*l*. 12*s*. 6*d*., and that the corps has in store rifles sufficient to arm seventy men. It has been further suggested that subscriptions might be received from the public, and applied in cases where Volunteers were unable to pay more than half the necessary expense.

It may not be generally known that a Corps, analogous in some respects to the one already described, has existed since the time of Edward III. in the Channel Islands, where the gallant and loyal inhabitants have for centuries banded themselves together for the defence of their hearths and homes. This militia resembles a *land-wehr* rather than a purely volunteer force, in the ordinary acceptation of the term, and comprises every male inhabitant, between the ages of fourteen and sixty, capable of bearing arms, be his rank what it may. It is an Anglo-Norman institution, whereby gratuitous suit and service is performed as an equivalent for the enjoyment of certain ancient privileges.

As soon as a lad attains the proper age, his name is enrolled on the militia-list, he is regularly drilled, and in due time is clothed, armed, and attached to one of the battalions belonging to his district. He continues to serve to the end of a prescribed period, unless exempted by a medical board.

The force in Jersey and Guernsey consists of battalions of artillery and infantry, which are called out for parades and field-days at the command of the Lieutenant-Governor, who bears the rank of a Major-General in her Majesty's service. Regular attendance, and perfect subordination when under arms, are secured by fines, and, in default of payment, by imprisonment. The duties discharged by these troops during the French revolutionary war, in addition to frequent drills, were, statute labour for the repair of the sea-coast defences, and watch and ward by day and night. At this period Sir John Doyle was Lieutenant-Governor of Guernsey, and General Don, Governor of Jersey; and by both of those distinguished officers the efficiency of the militia, and their soldier-like bearing, were highly appreciated and publicly acknowledged.

The only expense entailed upon Government for the maintenance of these troops is, the payment of a certain number of Adjutants, or inspectors, providing the privates and non-commissioned officers with arms, and ammunition, and clothing them in the British uniform at stated, but distant periods. Those who are incapable of bearing arms are appointed to warn the rest, and this arrangement is so efficient that the whole force can be concentrated on any one point at all times with very little delay. As a proof of this, it may be mentioned, that when her Majesty visited Guernsey in August, 1846, her arrival was not notified until late in the evening, and by daybreak the following morning, each soldier appeared at the rendezvous, armed and in uniform.

The infantry, besides being instructed in battalion drill, are now trained for rifle practice, and it is more

VOLUNTEER CORPS DISTINCT FROM CLUBS. 61

than probable that a very efficient force of marksmen will in process of time be organized; so that in the event of a rupture with France their skill as shots would be as formidable as that of the Vendeans or the Tyrolese. That they can be relied upon in cases of emergency is matter of history, since in 1781 the Jersey Militia supported the regulars in repelling a French invasion when De Ruttecourt landed there; and in 1783 the Guernsey Militia aided in quelling a mutiny which had broken out in an Irish regiment quartered in the citadel.

It would be no inconsiderable advantage to this country, if the principle here detailed were extended to all her colonies. Sir E. B. Lytton recommended a proposition to this effect, no longer ago than last year, strongly advocating the formation of Colonial Volunteer Corps, much upon this model.

To revert, however, to the question more immediately before us, we will suppose that it is intended to organize a Regiment of Volunteers, and that a sufficient number to begin with—say 300 names—have been subscribed.

Two classes of members should be nominated: those who from various circumstances can contribute to the funds only, without giving personal attendance; and those who are willing to serve, to arm and equip themselves, and to pay besides a moderate yearly subscription. Of course, if the Legislature does its duty, we may shortly add a third class, who being provided with arms, &c., by the State, and receiving some reasonable remuneration, will give their service only.*

* As matters now stand, however, all that the Government has yet done is to vouchsafe permission to those who like to put themselves to the expense of 10*l.* or 20*l.* to apply to a Lord-Lieutenant for leave to do so.

As in the formation of a Club, so in the present instance, the selection of a ground, and the erection of two or three different targets, with a sufficient mound, or wall behind, is the first consideration. As regards the Staff, differing as I do on this point from Mr. Drummond, I must say that no departure can safely be made from the detailed establishment of the VICTORIA RIFLES.

The Colonel, and other officers, may be unpaid, and the full number need not be filled up at once. Indeed, commissions, as already stated, should be held out as prizes for the more meritorious and the most proficient of the privates. To begin with, an active Adjutant is absolutely essential: a retired serjeant-major, or a half-pay Lieutenant in a rifle regiment, who would be content with a moderate remuneration for his services, would be the most useful man for this duty. A paid serjeant-armourer, a bugler, and marker, must also be engaged, and if the funds will admit of it, a drill-serjeant, through whose hands each recruit should pass as he enters; nor should any be permitted to take their places in the ranks till their proficiency in squad drill has been attested by the instructor. They should then go through a complete course of musketry, and "judging distance drill," as taught at Hythe. All that is requisite to be known of this, will be found in "THE RIFLE AND HOW TO USE IT," 4th edition, p. 143—157; as also in another part of that work, how to select a rifle, together with tolerably minute instructions for the tyro in target practice.

The recruits will now be prepared to receive instruction in the manual and platoon exercises, or if they can spare the time, these instructions can be imparted concurrently with those already indicated.

SECTION IV.

MANUAL AND PLATOON EXERCISES.

Preliminary Observations.

THE following instructions are strictly in accordance with the new drill, as now practised by all British Light Infantry corps.

It is a great error to imagine that proficiency at the target, is all that is required to constitute an efficient Volunteer. It is absolutely essential that he should, in addition, act with precision in concert with others; and learn that implicit obedience to his officers without which no man has any pretension to be considered a soldier, for without it he would be worse than useless in the field.

Any body of men, whatever their skill as shots might be, who should venture upon active service, under the supposition that each individual would there be free to act independently, as might seem best to himself, would soon find, though too late, that their zeal and valour would be wholly misplaced, that they would be continually in the way, without much chance of doing good, and would, perhaps, not unfrequently suffer as severely from the fire of their friends as of their foes.

Rifle clubs, and rifle associations, are invaluable institutions in their way, but their members must not imagine that, even after months of mere practice at the target, they can be deemed useful soldiers.

They have, it is true, learnt an important branch of their duty, but they are then only fit to be drafted into a Rifle Corps. In fact, until they are so enrolled, and have taken the prescribed oath, all drilling and training is illegal, and subjects those concerned in it to penal consequences.

All drill for Volunteers should be as simple, and as little irksome as possible. The chief objects are :—

1st. That each individual should have sufficient knowledge of every part of his rifle to enable him to take it to pieces, and put it together again, when requisite.

2nd. That he should know how to load it properly.

3rd. How to regulate his aim according to the distance of the object to be hit.

4th. Be practised in estimating distances within the ordinary range of his rifle.

5th. Be able on all occasions to take up a position in which he will be enabled—

To aim with facility;
To keep his body steady without constraint;
To be careful above all not to allow his sights to incline on one side or the other;
To support the recoil.

6th. When pulling or rather "pinching" the trigger in the act of firing, to be particularly careful not to derange his aim.

These few simple rules comprise nearly all that is really necessary to enable any man to attain the maximum effect with his rifle.

The author has purposely avoided encumbering the present work by introducing the ordinary recruit or squad drill, but as it is essential that Volunteers, when called out, or co-operating with other troops, should be perfectly acquainted with the Light Infantry drill,* the details of that exercise have been fully given.

That drill has of late been so much modified, that every existing manual may be considered obsolete. The author has endeavoured still further to simplify the instructions contained in the following pages, so as to render them intelligible to every capacity.

The drill here described has reference to the short regulation rifle. The introduction of breech-loaders would at once render unnecessary many of the movements now requisite with the more clumsy and cumbersome muzzle loader.

---

* The following observations on this subject by the late Sir Charles James Napier well merit introduction here :—

"With regard to Volunteer Corps, they should consist of from one to four companies, of 100 men each, with a captain and two lieutenants. Let each man carry two cartridge boxes, made to slide on a girdle round the waist, so that one may be carried before and one behind, each holding thirty rounds of ammunition : thus the weight would be divided.

"Get some old soldier for your adjutant, to teach you, not a long course of drill, but just seven things, viz. :—

"1. To face right and left by word of command.
"2. To march in line and in column.
"3. To extend and close files as light infantry, with 'supports.'
"4. To change front in extended and in close order.
"5. To relieve the skirmishers.
"6. To form solid squares and 'rallying squares.'
"7. To form an advanced guard.

"These seven things are all that you require; do not let any one persuade you to learn more.

"Let the target practice be constant. Also habituate your corps to take long marches of from fifteen to twenty miles, with arms and ammunition on; and also in running, or what is called 'double quick time.' These must be arrived at by gradually increasing from small distances. No single man, much less a body of men, can make these exertions without training. Also subscribe for premiums to those who are the best shots. Do not be exclusive in forming your corps; take your gamekeepers as your comrades, and any of your labourers that will enrol themselves. A gentleman will find no braver or better comrades than among his own immediate neighbours and tenants. Should you require to throw up a breastwork, they will be more handy with the spades and pickaxes than yourselves."

SECTION V.

THE NEW MANUAL EXERCISE FOR RIFLEMEN.

At the word "*Shoulder Arms*," the rifle is to be carried in the right hand at the full extent of the arm, close to the side; guard to the front, the forefinger and thumb round it, the remaining fingers under the cock, the upper part of the barrel close in to the hollow of the shoulder.

1st.
Secure,
Arms.
 — Seize the rifle with the left hand at the lower band, raising it a few inches by slightly bending the right arm, but without moving the barrel from the shoulder, and slip the thumb of the right hand under the cock, bringing the fingers under the guard to the front, and slanting downwards; both arms close to the body; left hand square with the left elbow.

Two. — Pass the rifle smartly to the left side, and cant the butt to the left rear, with the right hand to bring the rifle under the arm, quitting the right hand immediately to the right side; the cock to be close up under the arm-pit, the barrel to be uppermost, slanting downwards and inclining to the right front; the rifle to be firmly grasped with the left hand, which is to be rather below the hip; the left elbow a little to the rear; the lock not to be visible.

N.B.—In marching or standing at ease, the right hand is to grasp the rifle above the lower band, the stock to rest on the left arm, and the left hand to lay hold of the right arm above the wrist.

Secure Arms.   Shoulder Arms.   Order Arms.   Port Arms.   Present Arms.

2nd. Shoulder, Arms.	Carry the rifle to the right side with the left hand, and seize it with the forefinger and thumb of the right hand round the guard (remaining fingers under the cock), at the full extent of the arm without constraint, the left hand to steady it in the shoulder; arm close to the body.
Two.	Bring the left hand smartly to the left side.
3rd. Order, Arms.	Seize the rifle with the left hand, thumb and fingers round the piece, the little finger in line with the point of the right shoulder, but without moving the barrel therefrom, arm close in to the body.
Two.	Bring the rifle down in the left hand as low as the left arm will admit, keeping the arm and rifle close to the body, and with the right hand, which is to seize the rifle between the bands, place the butt quietly on the ground, even with the toe of the right foot; bringing the left hand at the same instant smartly to the left side; the right arm to be slightly bent, the thumb pressed against the thigh, fingers slanting towards the ground.
4th. Fix, Swords.	Place the rifle with the right hand smartly between the knees, guard to the front, and immediately seize the handle of the sword with the right hand (the left hand holding the scabbard), and draw it towards the muzzle, raising the point upwards when clear of the scabbard, and seize the rifle with the left hand at the "nose cap," then place the back part of the handle against the lock side of barrel, knuckles to the right, arm close to the body, and slide the spring on to the catch, and the ring on to the muzzle; when this is done, seize the rifle with the right hand between the bands, bring the left hand smartly to the left side, and come to the position of "*Order Arms.*"

MANUAL EXERCISE. 69

5th. *Shoulder Arms.*	Raise the rifle with a smart cant of the right hand, and seize the trigger guard between the forefinger and thumb at the full extent of the arm, the remaining fingers under the cock, at the same time seize the rifle also with the left hand in line with the elbow to steady it in the shoulder, arm close to the body.
Two.	Bring the left hand smartly to the left side.
6th. *Present, Arms.*	Seize the rifle with the left hand as in the first motion of " *Secure Arms.*"
Two.	Raise the rifle with the right hand perpendicularly from the shoulder to the *poise*, bringing it in front of the centre of the body, lock to the front, at the same time place the left hand smartly on the stock with the fingers pointing upwards, the thumb close to the forefinger, the point in line with the mouth, the wrist on the trigger guard, the left elbow close to the butt, the right elbow and butt close to the body.
Three.	Bring the rifle down with a quick motion as low as the right hand will admit without constraint, guard to the front, and grasp it with the left hand, the little finger touching the projection above the lock plate, thumb between stock and barrel, at the same time draw back the right foot so that the hollow of it may touch the left heel; lightly holding the small of the butt with the right hand, fingers pointing rather downwards; the rifle in this position to be totally supported with the left hand, close in front of and opposite the centre of the body.
7th. *Shoulder, Arms.*	Bring the rifle to the right side, and seize it with the right hand, the thumb and forefinger round the guard at the full extent of the arm, remaining fingers under the cock, bringing the left hand square with the left elbow and the right foot to its original position, both arms close to the body.
Two.	Bring the left hand smartly to the left side.
8th. Port, Arms.	Seize the rifle with the left hand, as in the first motion of " *Secure Arms.*"
Two.	Bring the rifle in the left hand to a diagonal

position across the body, lock to the front, and seize the small of the butt at once with the right hand, thumb and fingers round the stock, the left wrist to be opposite the left breast, both elbows close to the body, the muzzle slanting upwards, so that the barrel may cross opposite the point of the left shoulder.

9th.
Charge,
Swords.

Make a right half face by raising the toes and turning upon the heels, the right toes to point to the right, the left full to the front, and bring down the rifle to nearly a horizontal position at the right side, with the muzzle inclining a little upwards, the right wrist to rest against the hollow of the thigh below the hip, the thumb of right hand pointing to the muzzle.

When a battalion in line charges with swords, the whole advance at a firm quick step with shouldered arms; at the command "*Prepare to charge,*" the rifles of the front rank will be brought to the long trail, and those of the rear rank to the slope; at the word "*Charge,*" the rifles of the front rank will be brought smartly to the charging position, and the pace increased to "*double march,*" carefully avoiding too much hurry. At the word "*Halt,*" both ranks halt and shoulder arms.

10th.
Shoulder,
Arms.

Raise the rifle perpendicularly at the right side, seizing it with the right hand, the thumb and forefinger round the guard at the full extent of the arm, the remaining fingers under the cock, the left hand holding the piece above the lower band square with the left elbow, both arms close in to the body, facing at the same time to the front.

Two.

Bring the left hand smartly to the left side.

N.B. In the foregoing motions care must be taken to preserve the squareness of the body, and to avoid raising or sinking the shoulder.

11th.
Slope,
Arms.
Two.

Seize the rifle with the left hand as in the first motion of "*Secure Arms.*"

Bring the rifle to the left side, and seize it at once with the left hand, the two first joints of the fingers grasping the butt, the thumb alone to be

MANUAL EXERCISE. 71

	on the heel of it, the muzzle to slant to the rear, and the guard pressed gently against the hollow of the shoulder; the upper part of the left arm to be close in to the side, the hand in a line with the elbow, the toe of the butt opposite the centre of the left thigh; the right hand holding the small, thumb and fingers round the stock, arm close in to the body.
Three.	Bring the right hand smartly to the right side.
12*th.* *Stand, at ease.*	Bring the right hand smartly across the body and place it on the left hand; thumbs of both hands on the heel of the butt, that of the left nearest to the lock; at the same time move the left foot six inches to the front, with the toes pointing to the left front (feet separated), the left knee to be slightly bent, the greater part of the weight of the body to be brought on the right leg; no constraint.
13*th.* *Attention.*	Bring the left foot, back in line with the right, and the right hand smartly to the right side.
14*th.* *Carry, Arms.*	Seize the small of the butt with the right hand, fingers and thumb round the stock, arm close in to the body.
Two.	Bring the rifle to the right side, and seize it with the forefinger and thumb round the guard, the remaining fingers under the cock, at the full extent of the right arm; the left hand to be brought across the body with the fingers extended on the rifle in line with the elbow to steady it to the shoulder; thumb close to the forefinger.
Three.	Bring the left hand smartly to the left side.
15*th.* *Order, Arms.*	As detailed in No. 3.
16*th.* *Unfix, Swords.*	Place the rifle with the right hand smartly between the knees, guard to the front, and seize it with the left hand at the "nose cap," as also the handle of the sword with the right hand, knuckles to the front, fingers pointing downwards; then

with the forefinger press the spring inwards, raise the sword upwards, and drop the point, with the edge to the front, towards the scabbard, raising the right elbow in doing so, at the same time move the left hand smartly and seize the scabbard to guide the sword into it, this being done, seize the rifle with the right hand, and come to the position of "*order arms*."

17*th.*
Stand,
at ease.
Push the muzzle of the rifle to the front with the right hand, arm close to the side; at the same time move the left foot six inches to the front, &c., as detailed at No. 12.

The foregoing exercise is applicable for serjeants, and supersedes the "Fusil exercise."

Serjeants in line will remain steady at "*shoulder arms,*" during the performance of the manual and platoon exercise.

In taking up an alignment, the rifle is to be raised with the left hand grasping the butt, thumb on the heel-plate, the small of the butt being seized with the right hand, the piece brought before the body, the barrel to the front.

Time.

In the manual exercise the different motions should be performed, with one pause of the slow time of march, between each, except that of *fixing swords*, in which longer time must be given; one pause should also be made between the first and last parts of the words of command, for instance, *shoulder* (one pause) *arms*, both in the manual and platoon.

The manual is not to be executed by one word, but each separate command is to be given by the officer who commands the body performing it.

In ordering arms from any position, great care must be taken to prevent the rifle from striking the ground, nor is it at any time to be unnecessarily shaken.

MANUAL EXERCISE. 73

	From the "Order" with Long and Short Rifle.	From the Shoulder with Long Rifle.	From the Advance, or Shoulder with Short Rifle.
Trail, Arms.	Bring the rifle to a horizontal position at the right side, holding it with the right hand behind the lower band (thumb and fingers round the piece) at the full extent of the arm; at the same time the rear rank man will take a short pace to the rear, so that the muzzle of his rifle may be just in front and clear of the wrist of his front rank man.	Seize the rifle with the right hand under the lower band, arm close in to the body, thumb and fingers round the piece. *Two.*—Bring the rifle in the right hand to a horizontal position at the right side, and hold it at the full extent of the arm; at the same time the rear rank man will take a short pace to the rear, so that the muzzle of his rifle may be just in front and clear of the wrist of his front rank man.	Seize the rifle with the left hand, little finger in line with the elbow, arm close in to the body. *Two.*—Bring the rifle to a horizontal position at the right side, seizing it at once with the right hand behind lower band (thumb and fingers round the piece) at the full extent of the arm, bringing the left hand smartly to the left side; at the same time the rear rank man will take a short pace to the rear.

Change, Arms. { Bring the rifle to a perpendicular position at the right side, and seize it with the left hand close above the sight, and carry it round to the left side, bringing it to a horizontal position at the full extent of the arm.

When moving with *trailed arms*, at the word "*halt,*" arms are to be ordered, when the rear rank will close to the front. In Rifle corps at the word "*March,*" arms are to be trailed without any command.

Order, Arms (from the trail.) { Bring the rifle at once to a perpendicular position at the right side, and place the butt quietly on the ground, and come to the position of "*order arms.*"

N.B.—To relieve the men, the arms may be "trailed" on the line of march, or when moving as light infantry.

Trailed arms must never be used with fixed bayonets, except in preparing to charge; nor in file marching, nor in any field movements where close marching is required.

If required to move a few paces backwards or forwards when at *ordered arms*, the rifle is merely to be raised from the ground, keeping the barrel close to the shoulder.

TO PILE ARMS.

The company standing in close order, with ordered arms, are to be numbered from right to left.

Pile, Arms.
> At the word "*pile*," the rear rank step ten inches to the rear, and the front rank draw back their right feet in order to face to the right about; at the word "*arms*," the front rank will face about, bringing their rifles with them to ordered arms; the front and rear rank men will then place the butts, locks inwards, against the inside of their outer feet as close to the heel as possible, after which the right file rear rank and the left file front rank will incline their rifles towards each other, and cross ramrods; the right file front rank will at once place his left hand round the muzzle of his left file, bearing it from him, and with his right hand lock ramrods by passing his between the ramrods and to the right of the muzzles of the other rifles, the left file rear rank will then lodge his rifle between the muzzles of the rifles of the front rank, sling uppermost. When there is an odd file, the front and rear rank man will lodge his rifle against the pile nearest his right hand.

Stand, clear.
> Ranks take a pace of ten inches backwards and face towards the pivot flank.

Stand to.
> Ranks facing towards the pivot flank, will face inwards and close on their arms by taking a pace of ten inches forward.

Unpile, Arms.
> At the word "*unpile*," seize the rifle with the right hand under the top band, front rank at the same time drawing back their right feet in order to face to the right about; at the word "*arms*," unlock the ramrods without hurry, by inclining the butts inwards, and come to ordered arms, the front rank will then "*front*," and the rear rank close on it by taking a pace of ten inches forward.

N.B.—It is necessary to be careful in piling and unpiling arms to prevent damage being done to the ramrods and sights.

THE PLATOON EXERCISE.

The young rifleman being now supposed to possess a competent knowledge of the several motions detailed in the MANUAL EXERCISE, should be taught the PLATOON EXERCISE. For this purpose a squad of eight or ten men will be formed in a single rank at close files:—

    1st. To load and fire standing.

    2nd. To load and fire as a front rank kneeling.

    3rd. To load and fire as a rear rank kneeling.

WORDS OF COMMAND.	EXPLANATIONS.
Platoon Exercise by motions, standing.	To serve as a caution.

	From Shouldered Arms.	From Advanced Arms, or Shoulder Arms, with short Rifles.	From Ordered Arms.
Prepare, to Load.	Seize the rifle with the right hand as in the first motion of *order arms*, at the same time make a right half-face by raising the toes and turning upon the heels, left toes to point to the front, right toes to the right, eyes to the front.	Seize the rifle with the left hand, the little finger in line with the right shoulder, left arm close in to the body, at the same time make a right half-face by raising the toes and turning upon the heels, the left toes to point to the front, the right toes to the right, eyes to the front.	Make a right half-face by raising the toes and turning upon the heels, the left toes to point to the front, the right to the right, carrying the rifle round with the body, eyes to the front.
Two.	*As a front rank*, carry the left foot ten inches to the left front (viz. 6 to the front and 8 to the left), moving the body with it; *as a rear rank*, carry the left foot six inches to the front, moving the body with it also, toes in both cases to point direct to the front; at the same time square the shoulders to the front from the hip, and bring the rifle down perpendicularly in the right hand opposite the left breast, to the full extent of the arm, and seize it with the left hand at the nose cap, thumb and fingers round the stock and barrel, and place the butt without noise on the ground, close against the inside of the left foot, the heel of it to be in a line with the ball of the big toe, barrel to the front and perpendicular, left arm close to the side; carrying the right hand at once to the pouch, (elbow to the rear,) and take up a cartridge, holding it with the fore-finger and thumb close to the top, with the bullet in the palm of the hand.	*Two—as a front rank*, carry the left foot ten inches to the left front, (viz., 6 to the front and 8 to the left,) moving the body with it; *as a rear rank* carry the left foot six inches to the front, moving the body with it also, at the same time square the shoulders to the front from the hip, and bring the rifle down in the left hand, and place the butt without noise on the ground close against the inside of the left foot, the heel of it to be in a line with the ball of the big toe, front and perpendicular, and slip the left hand smartly to the nose cap, thumb and fingers round the stock and barrel, left arm close to the side; carrying the right hand at once to the pouch (elbow to the rear), and take up a cartridge, holding it with the fore-finger and thumb close to the top, with the bullet in the palm of the hand.	*Two—as a front rank*, carry the left foot ten inches to the left front, (viz., 6 to the front and 8 to the left,) moving the body with it; *as a rear rank* carry the left foot six inches to the front, moving the body with it also, and at the same time square the shoulders to the front from the hip, and pass the rifle smartly to the left hand, which will seize it at the "nose-cap," thumb and finger round the stock, and place the butt without noise on the ground, close against the inside of the left foot, the heel of it to be in a line with the ball of the big toe, barrel to the front and perpendicular, left arm close to the side, carrying the right hand at once to the pouch (elbow to the rear), and take up a cartridge, holding it with the fore-finger and thumb close to the top, with the bullet in the palm of the hand.

N.B. The feet as above detailed, being at right angles, care must be taken that this angle is not increased by turning the right toes to the rear, as such would tend to alter the proper and essential position of the right shoulder in loading and firing.

PLATOON EXERCISE.

Load. — Bring the cartridge to the forefinger and thumb of left hand, and with the arm close in to the body, tear off the end of it carefully so as not to lose any of the powder; any motion which may be necessary to be from the wrist only.

Two. — Bring the cartridge to the muzzle of the rifle, and pour the powder into the barrel, inclining the palm of the hand to the front, and bringing the right elbow square with the wrist in doing so, the thumb of left hand to point to the muzzle.

Three. — Reverse the cartridge by dropping the hand over the muzzle, bringing the fingers round the barrel, knuckles to the front, and put the bullet into the barrel nearly as far as the top, holding the paper above the point of the bullet between the forefinger and thumb, still keeping the right elbow square with the wrist.

Four. — By a turn of the wrist from left to right, pressing the little finger against the barrel, and dropping the right elbow into the side, tear off the paper which is held between the forefinger and thumb; when this motion is completed, the little finger should rest against the side of the barrel, the knuckles inclined towards the ground.

Five. — Seize the head of the ramrod between the second joint of forefinger and thumb, knuckles towards the body.

Rod. — Draw the ramrod smartly half out of the stock; seize it exactly in the middle between the first two fingers and thumb of the right hand, the forefinger to be in a line with the muzzle of the rifle, knuckles towards the body, the remaining fingers closed in the hand, the elbow square with the wrist; the thumb of left hand pointing to the muzzle.

Two. — Draw the ramrod entirely out with a straight arm, turn it (dropping the head to the front, the point to pass close by the side of the left ear), and place it on the top of the bullet; the ramrod to be perpendicular, and held in the middle between the first two fingers and thumb of the right hand,

78   RIFLE VOLUNTEERS.

the remaining fingers closed in the palm of it, the knuckles full to the front, the arm to be kept as close to the ramrod as possible without constraint, and without altering the squareness of the shoulders.

Home. Force the bullet straight down the barrel until the second finger touches the muzzle of the rifle, bringing the elbow down close in to the body at the same time, inclining the knuckles to the right.

Two. Move the right hand smartly to the point of the ramrod, and seize it between the first two fingers and thumb, the remaining fingers to be closed in the hand, the knuckles full to the front, the arm to be kept as close to the ramrod as possible without constraint, and without altering the squareness of the shoulders.

Three. Force the bullet steadily straight down to the bottom, bringing the elbow down close in to the body at the same time, inclining the knuckles to the right.

Four. By two steady and firm pressures (raising the ramrod about one inch on each occasion) ascertain that the bullet is resting on the powder; all strokes which may indent the point of the bullet to be avoided.

Return. Draw the ramrod smartly half out of the barrel, and seize it in the middle between the first two fingers and thumb of the right hand, the forefinger in a line with the muzzle of the rifle, knuckles towards the body, the remaining fingers closed in the hand, the elbow square with the wrist.

Two. Draw the ramrod entirely out with a straight arm, turn it (dropping the point to the front, the head to pass close by the side of the left ear), and put it into its place at once, pressing the ramrod towards the body in doing so, to prevent the point catching the band or otherwise doing injury to the stock; move the right hand smartly at the same time, and place the second joint of the forefinger

PLATOON EXERCISE. 79

(the remaining fingers to be closed in the hand) on the head of the ramrod and force it home, then seize it between the second joint of forefinger and thumb, and drop the left hand smartly at the same instant to its full extent, and seize the rifle; the arm to be close in to the body.

N.B.—In performing the motions of "*Rod*" and "*Return*," care must be taken that the ramrod rubs as little as possible against the sides of the barrel or muzzle, that the shoulders are preserved square to the front, and that the body is kept perfectly steady.

Cap. Let the shoulders resume the half-face, and bring the rifle to a horizontal position at the right side with the left hand, which is to grasp it firmly behind the lower band, but not nearer to the nipple than the projection in front of the lock-plate against which the little finger may rest, the thumb between stock and barrel, the left arm to be close in to the body as a support, at the same time meet the "small of the butt" with the right hand, elbow to the rear, and hold it lightly with the fingers *behind the trigger guard*, and half cock the rifle, the thumb to remain on the cock; *as a front rank* the "small of the butt" to be pressed against the hip, *as a rear rank* four inches above it.

Two. Advance the fingers to the nipple, and with the forefinger throw off the old cap.

Three. Carry the hand to the cap pocket, and take up a cap between the forefinger and thumb, the remaining fingers to be closed in the hand, elbow to the rear.

Four. Put the cap straight upon the nipple, looking to the front after doing so.

Five. Press the cap home with the flat part of the thumb, with the fingers closed in the hand and against the lock-plate.

Six. Bring the hand to the "small" of the butt, and hold it lightly with the fingers *behind the trigger guard*, thumb pointing to the muzzle.

As a Front (or Rear) Rank at — yds. Ready.	From Shouldered Arms.	From Advanced Arms, or Shouldered Arms with short Rifles.	From Ordered Arms.
Carry the right hand to the sight, and with the fore-finger and thumb adjust the sliding bar, placing the top even with the line, or to the place that indicates the elevation necessary for the distance named, then raise the flap without a jerk from the top if required, after which bring the hand back to the small of the butt, and full cock the rifle, and hold it lightly with the fingers behind the trigger guard, thumb pointing to the muzzle, and fix the eye STEADFASTLY on some object in front.	Make a right half-face by raising the toes and turning upon the heels, the left foot to point direct to the front, the right foot to the right, at the same time seize the rifle with the right hand at the small of the butt, thumb pointing to the muzzle. Two.—Bring the rifle to a horizontal position at the right side, meeting it with the left hand, which is to grasp the stock firmly behind the lower band, but not nearer to the nipple than the projection in front of the lock plate, against which the little finger may rest, thumb between stock and barrel, the left arm to be close in to the body as a support; the small of the butt, as a front rank, pressed against the hip, as a rear rank, four inches above it, as a front rank carry the left foot ten inches to the left front (viz., six inches to the front and eight to the left) moving the body with it, as a rear rank carry the left foot six inches to the front, moving the body with it also, toes in both cases to point direct to the front, and proceed as detailed in the left-hand column.	Make a right half-face by raising the toes and turning upon the heels, the left foot to point to the front, the right foot to the right, at the same time seize the rifle with the left hand, the little finger in line with the right elbow. Two.—Bring the rifle to a horizontal position at the right side, grasping it with the left hand firmly behind the lower band, but not nearer to the nipple than the projection in front of the lock plate, against which the little finger may rest, the thumb between stock and barrel, the left arm close in to the body as a support; the small of the butt as a front rank pressed against the hip, as a rear rank, four inches above it, then carry the left foot as a front rank ten inches to the left front (viz., six to the front and eight to the left), moving the body with it, as a rear rank carry the left foot six inches to the front, moving the body with it also; toes in both cases to point direct to the front, and proceed as detailed in the first column.	Make a right half-face by raising the toes, &c., as before directed, carrying the rifle round with the body, and place the thumb of the right hand smartly behind the barrel and seize the rifle. Two.—Bring the rifle to a horizontal position at the right side, grasping it with the left hand, as explained in the adjoining columns, &c., then proceed as detailed in the first column.

PLATOON EXERCISE.

Present. — Raise the rifle to the shoulder at once, carrying it to the front so as to clear the body in doing so, but without moving the left hand from the place at which it grasps the stock at the capping position, or stooping the body, or raising the heels off the ground (the rifle resting solidly in the palm of the left hand), at the same time raise the right elbow nearly square with the right shoulder, but inclined to the front of it, so as to form a bed for the butt, the centre of which press firmly to the shoulder with the left hand, and bring the left elbow well under the rifle to form a support; the right hand to lightly hold the small of the butt, the thumb pointing to the muzzle, which is to be a few inches below the object the right eye is fixed upon, the forefinger along the outside of the trigger guard, and the left eye closed: the arm of the front rank man is not to be raised too high, as he will thereby prevent his rear rank man taking aim.

N.B.—As the recruit will not get into the position here detailed without much practice, the instructor will frequently command "as you were," (when the rifle is to be brought down to the right side), and point out the defects observed; by this means the recruit will soon get into the position readily, acquire a full command of his rifle with the left hand, and become habituated to handle it with expertness.

Two. — Place the forefinger round the trigger like a hook (that part of it between the first and second joint to rest flat on the trigger), inflate the lungs fully (this is of the utmost importance), and restrain the breathing.

Three. — Raise the muzzle steadily, until the top of the foresight is brought in a line with the object through the bottom of the notch of the back sight.

Four. — Pinch the trigger without the least jerk or motion of the hand, eye, or arm, until the cock falls upon the nipple, *keeping the eye still firmly fixed upon the object for at least three seconds.*

Five. Bring the rifle down to the capping position, and shut down the flap, and immediately seize the rifle with the right hand close in front of the left, fore arm close to the barrel; and after a pause of the slow time, taking the time from the right, turn the barrel at once downwards, and bring the rifle to a perpendicular position in the right hand, and come to the position of "prepare to load," second motion.

NOTE.—Too much pains cannot be taken to insure that the recruit takes a deliberate aim at some *positive object* whenever he brings the rifle to the " Present."

In the "Present," the body is to be firm and upright, the butt to be pressed firmly into the hollow of the shoulder,[*] so as to avoid the kick from the recoil; the rifle to rest solidly in the left hand, firmly grasped, but without rigidity of muscle, the sight to be upright, and in aiming, the muzzle to be steadily raised until the top of the foresight is aligned upon the object on which the right eye is fixed, through the bottom of the notch of the back sight, the left eye being closed, and the breathing restrained. In firing, the trigger is to be pinched, by pressure alone, without any motion of the hand, eye, or elbow; the right eye to continue fixed on the object after snapping, to ascertain if the aim has been deranged in any way.

The position of the head with reference to the butt of the rifle when taking aim depends entirely on the distances fired at, or the elevation used. At short distances, the butt must be brought to the head by raising the shoulder, or the cheek so placed on the butt as to fix the eye on the object through the bottom of the back sight without too much stooping of the head: as the distances increase, the head must be raised or the shoulder lowered.

Load. As before detailed, by motions, until the recruit has attained such a knowledge of them as to be capable of combining them in regular order.

[*] See "THE RIFLE, AND HOW TO USE IT." Fourth edition, p. 130.

PLATOON EXERCISE. 83

Shoulder, Arms.
At the word "*shoulder*" bring the left foot back to the right, (placing the heel behind that of the right foot), and at the word "*arms*," face to the front by raising the toes and turning upon the heels, at the same time throw the rifle with the right hand on to the left shoulder, and grasp, at the full extent of the arm, the butt with the left hand, the fingers of the right hand to be under the cock and close to the lock side of stock, thumb pointing to the muzzle.

Two. Bring the right hand smartly to the right side.

From the capping-position.

Advance, Arms. or Shoulder, Arms. with short Rifle.
At the word "*advance*" or "*shoulder*" bring the left foot back to the right, (placing the heel behind that of the right foot), and at the word "*arms*" face to the front by raising the toes and turning upon the heels, at the same time bring the rifle to a perpendicular position at the right side with the left hand, fingers extended and in line with the elbow, and seize it with the forefinger and thumb of the right hand round the trigger-guard, the remaining fingers under the cock.

Two. Bring the left hand smartly to the left side.

Order, Arms.
At the word "*order*" bring the left foot back to the right (placing the heel behind that of the right foot), and seize the rifle with the right hand close in front of the left, forearm close to the barrel; at the word "*arms*" face to the front by raising the toes and turning upon the heels, and with the right hand place the butt quietly on the ground at the right side, even with the toe of the right foot, &c., as detailed in the *manual exercise*.

G 2

To load and fire kneeling.

Platoon Exercise by Motions, as a front (or rear) rank kneeling. To serve as a caution.

	From Shouldered Arms.	From Advanced Arms, or from Shouldered Arms with short rifles.
Prepare to Load.	Seize the rifle with the right hand under the cock, as detailed in the first motion of "*secure arms*," and at the same time make a right half-face by raising the toes and turning upon the heels, the left toes to point to the front, right toes to the right.	Seize the rifle with the left hand, the little finger in line with the right elbow, keeping the left arm close in to the body; at the same time make a right half-face, by raising the toes and turning upon the heels, the left toes to point to the front, the right toes to the right.
Two.	Grasp the rifle with the left hand, the little finger as high as the shoulder, the elbow close in to the lock plate; at the same time carry the right foot twelve inches to the rear, and place the toe of the boot on the ground as much to the left of the left heel as will bring the right knee of the *front rank* six inches to the right when on the ground, and that of the *rear rank* twelve inches to the right; the foot to be nearly perpendicular, the left leg straight.	*Two.*—Carry the right foot twelve inches to the rear, and place the toe of the boot on the ground as much to the left of the left heel as will bring the right knee of the *front rank* six inches to the right when on the ground, and that of the *rear rank* twelve inches to the right; the foot to be nearly perpendicular, the left leg straight.

PLATOON EXERCISE.

	Front Rank.	Rear Rank.
Three.	Sink down at once on the right knee, six inches to the right and twelve inches to the rear of the left heel, and square with the foot, which is to be under the body, and upright, the left leg to be as perpendicular as possible; at the same time bring the rifle down in the left hand, close in to the body, and pass the butt to the left rear over the right heel to the extent of the left arm, sling upwards, meeting the barrel with the right hand, the thumb in a line with the muzzle, the right arm to be close in to the body, the hand in front of the left breast, the shoulders to be brought nearly square to the front.	*Three.*—Sink down at once on the right knee, twelve inches to the right, and twelve inches to the rear of the left heel, and square with the foot, which is to be under the body, and upright; bringing the body nearly to the right about three-quarters face in doing so (the left leg inclining to the right); and at the same time carry the rifle in the left hand, and place the butt flat on the ground (lock uppermost), under the shin of the right leg of the front rank man of the file on the right, meeting the barrel with the right hand, the thumb in a line with the muzzle, the right arm close in to the side, the muzzle of the rifle as high, and in a line with the right shoulder, eyes to the right rear.
Four.	Seize the rifle with the left hand under the top swivel; the elbow to be close in to the left side, hand close under the left breast, the rifle close in to the hollow of the left side and as upright as possible; at the same time carry the hand to the pouch and take up a cartridge, holding it between the forefinger and thumb, close to the top, with the bullet in the palm of the hand.	*Four.*—Seize the rifle with the left hand under the top swivel, elbow close in to the body, hand in front of the right breast, with the bullet in the palm of the hand.

N.B.—As the length of leg, in very tall men, is greater than the breadth of body, it will be impossible, in close order, to get the knee square with the foot; in such cases, therefore, the knee is to be inclined to the front, but not beyond the inside of the right foot of right file.

Load. { In five motions, as detailed when loading standing; in seizing the head of the ramrod in the fifth motion, the front rank to incline the ramrod to the right to facilitate the drawing of it.

Rod. In two motions, as detailed when loading, standing.

Home. In four motions, as detailed, when loading standing.

Return. Draw the ramrod smartly half out of the barrel, and seize it in the middle, between the first two fingers and thumb of the right hand, the forefinger in a line with the muzzle, knuckles towards the body, the remaining fingers closed in the hand, the elbow square with the wrist.

Two. Draw the ramrod out with a straight arm, turn it (dropping the point towards the ground), replace it at once, pressing the ramrod towards the body, to prevent the point catching the band or doing injury to the stock; move the right hand smartly at the same time, and place the second joint of the forefinger (the remaining fingers to be closed in the hand) on the head of the ramrod and force it home; then seize it between the second joint of the forefinger and thumb, arm to be close in to the body, and slip the left hand to the full extent, and seize the rifle immediately below the lower band.

Cap.

Front Rank.	Rear Rank.
Let the body resume the right half-face, and with the left hand bring the rifle to a horizontal position at the right side, by raising the butt from the ground and passing it over the right heel, close to the body, and round in front of the left leg, and place the left forearm at once square on the left thigh six inches behind the knee; at the same time meet the small of the butt with the right hand, and hold it lightly with the *fingers behind the trigger guard*, and half cock the rifle, the thumb to remain on the cock;—the rifle to be grasped with the left hand as detailed when capping standing; the butt to be pressed against the side.	Let the body resume the right half-face, and with the left hand bring the rifle to a horizontal position at the right side muzzle to the front, and place the left forearm at once square on the left knee, at the same time meet the small of the butt with the right hand, and hold it lightly with the *fingers behind the guard*, and half cock the rifle, the thumb to remain on the cock;—the rifle to be grasped with the left hand, as detailed when capping standing, the butt to be pressed against the side.

Two.
Three.
Four. As detailed when capping standing.
Five. Six.

PLATOON EXERCISE.

	From Shouldered Arms.	From Advanced Arms, or Shouldered Arms, with Short Rifles.	From Ordered Arms.
As a front (or rear) Rank at — yards Ready.	Bring the weight of the body on to the right heel, then adjust the sight as before explained; after which bring the hand back to the small of the butt, and full cock the rifle, and hold it lightly, with the *fingers behind the guard*, thumb pointing to the muzzle, and fix the eye STEADFASTLY on an object in front.	As detailed in the first motion of "ready" from the shoulder standing. *Two.*—Bring the rifle to a horizontal position at the right side, as explained in the 2nd motion of "ready" from the *shoulder* standing, at the same time carry the right foot 12 inches to the rear, and place the toe of the boot as much to the left of the left heel as will bring the knee of the *front rank* six inches to the right when on the ground, and that of the *rear rank* 12 inches to the right; the foot to be nearly perpendicular, the left leg straight.	As detailed in the first motion of "ready," from advanced arms standing. *Two.*—Drop the rifle to a horizontal position at the right side, as explained in the 2nd motion of "ready" from *advanced arms* standing, at the same time carry the right foot 12 inches to the rear, and place the toe of the boot as much to the left of the left heel as will bring the knee of the *front rank* 6 inches to the right when on the ground, and that of the *rear rank*, 12 inches to the right; the foot to be nearly perpendicular, the left leg straight.	As detailed in the first motion of "ready" from ordered arms standing. *Two.*—Bring the rifle to a horizontal position at the right side, as explained in the second motion of "ready" from ordered arms standing, at the same time carry the right foot 12 inches to the rear, &c. &c., as explained in the adjoining column.

Three.—Sink at once on the right knee twelve inches to the rear, *as a front rank* six inches to the right, *as a rear rank* twelve inches to the right of the left heel, and square with the right foot, and bring the weight of the body on to the right heel; the left forearm to be placed on the left leg, and the butt pressed against the right side as when capping; then adjust the sight and cock the rifle, fixing the eye STEADFASTLY on an object in front; the thumb of the right hand to be placed on the stock pointing to the muzzle.

N.B.—When required to come to the "ready" kneeling, from the capping position standing, the left foot must be brought back to the right before sinking down on the right knee.

Present.	As detailed when coming to this position standing, without raising the body off the heel, and place the left elbow at once over the left knee to form a support. *N.B.* The note following the first motion of the *present standing* is applicable to this motion also.
Two.	
Three.	As detailed when firing standing.
Four.	

Front Rank.	Rear Rank.
Five. Bring the rifle down to the capping position, at the same time raise the body off the right heel, and place the left forearm square on the left thigh six inches behind the knee; then shut down the flap without a jerk, and return the hand to the small of the butt, count a pause of the slow time, and come to the position of "prepare to load" by carrying the rifle in the left hand, passing the butt round in front of the left leg close in to the body, to the left rear over the right heel, to the extent of the left arm, meeting it at the same time with the right hand, the thumb in line with the muzzle, then seize the rifle with the left hand under the top swivel, as detailed in the fourth motion of "*prepare to load*" *as a front rank kneeling*, &c.	Bring the rifle down to the capping position, at the same time raise the body off the right heel, and place left forearm square on the left knee; then shut down the flap without a jerk, and return the hand to the small of the butt, count a pause of the slow time, and come to the position of "prepare to load" by turning the rifle over in the left hand, and placing the butt on the ground, lock uppermost under the shin of the right leg of the front rank man of the file on the right, meeting the barrel with the right hand thumb in line with the muzzle, which is to be as high and in a line with the right shoulder, then seize the rifle with the left hand under the top swivel, &c., as detailed in the fourth motion of "*prepare to load*" *as a rear rank kneeling*.

N.B.—When required to load standing from the kneeling position:—After shutting down the flap, seize the rifle with the right hand close in front of the left, and rise to the half-face at the same instant, bringing the right heel before the left, still keeping the rifle in a horizontal position at the right side, then after counting a pause, taking the time from the right, turn the barrel at once downwards, and bring the rifle to a perpendicular position, and proceed as detailed in the second motion of "*prepare to load*" standing.

PLATOON EXERCISE. 89

Load. As before detailed, by motions, and so continue exercising until the recruit has attained such a proficiency as to be capable of combining the several motions in regular order.

Shoulder, Arms. At the word "*shoulder*" spring smartly to attention at the half-face, bringing the right heel in front of the left, still keeping the rifle in a horizontal position at the right side; at the word "*arms*" proceed as detailed, when coming to the shoulder from the capping position, standing.

Two. Bring the right hand smartly to the right side.

Advance, Arms, or Shoulder, Arms. with short Rifles. At the word "*advance*" or "*shoulder*" spring smartly to attention at the half-face, bringing the right heel in front of the left, still keeping the rifle in a horizontal position at the right side. At the word "*arms*," face to the front, &c., as detailed, when coming to the advance from the capping position, standing.

Two. Bring the left hand smartly to the left side.

Order, Arms. At the word "*order*" spring smartly to attention at the half-face, bringing the right heel in front of the left, still keeping the rifle in a horizontal position at the right side, and at the same time seize the rifle with the right hand close in front of the left, forearm close to the barrel; at the word "*arms*," face to the front, &c., as detailed, when coming to the order from the capping position, standing.

{From the capping position kneeling.}

The recruits, when thoroughly grounded in the foregoing instructions, may be practised in two ranks at close order in the different firings, as a company in line, as a wing of a battalion, firing a volley, and file firing.

For this purpose from twenty to thirty files, or a less number, are to be formed into two ranks at close order, with shouldered arms, fixed bayonets, and knapsacks on.

REVIEW EXERCISE.

Words of Command.	Explanation.
Platoon Exercise in slow time.	At this caution, the rear rank take a pace of nine inches to the front.
Prepare to Load.	In two motions, observing a pause of the slow time between each.
Load.	In five motions do. do. do.
Rod.	In two motions do. do. do.
Home.	In four motions do. do. do.
Return.	In two motions do. do. do.
Cap.	In six motions do. do. do.
Fire a Volley, at — yards. Ready.	One motion, the sight to be adjusted with care and without hurry.
Present.	In five motions. There is to be no hurry at any time in the performance of the third and fourth motions of the present.
In quick time, Load.	Perform the motions of loading smartly, but with the same correctness as if exercising in the slow time;—after returning the ramrods, the whole line to remain perfectly steady; after a pause of the slow time, taking the time from the right, come to the capping position, and proceed to cap, which must always be done after loading.
Shoulder, Arms.	As before detailed, the rear rank taking a short pace of nine inches to the rear when quitting the right hand.
Company, (Right Wing, or Battalion,) " Fire a Volley."	At this caution, the rear rank will take a pace of nine inches to the front.

PLATOON EXERCISE.

At — yds.
Ready. } As before detailed.

Present. As before detailed; after firing, make a pause (taking the time from the right), come down to the capping position, put down the flap, and immediately seize the rifle with the right hand close in front of the left, forearm close to the barrel, and after another pause, come to the position of "*Prepare to load*," and go on with the loading in the quick time without any command to do so.

Cease Firing. At the close of the "general," or at the command "cease firing," the company having completed its loading and capped, will receive the command "shoulder arms." If the company is at the ready when the "cease firing" sounds, it will be commanded to "half cock arms," to be performed as follows:—

Half-cock Arms. Place the thumb of the right hand on the comb of the cock and the forefinger on the trigger, and draw both back until the sear is disengaged from the "full bent of tumbler," then let the cock gently down (removing the forefinger from the trigger), and when it passes the "half bent," draw it back to half cock, after which put down the flap, and carry the right hand to the small of the butt, thumb pointing to the muzzle, *fingers behind the guard*.

When it is not intended to reload after firing, the command will be, "*Fire a Volley and shoulder*." "*At — yds. Ready*." After delivering the volley, make a pause, and taking the time from the right, come down to the capping position, shut down the flap, bring back the right hand to the "small of the butt," and in doing so close the heels, then after another pause, taking the time from the right also, come to the shoulder as before detailed.

NOTE.—When a column or line is required to load, the command is to be—

With Cartridge, or,
As with Cartridge. } As a caution, on which the rear rank will take a pace of nine inches to the front.

92   RIFLE VOLUNTEERS.

Load. { The loading to proceed in the quick time:— When in column, or when any person is immediately in front, the rifle, when brought to the capping position, is to be slanted with the muzzle inclining upwards, the flat part of the butt pressed against the thigh, in order to prevent the possibility of accident.

N.B.—When giving the command "Ready," some distance must always be named; should, however, the distance be omitted, the soldier must judge for himself the distance he is, from the object he is going to aim at, and adjust his sight accordingly.

File Firing.

File firing, from the right (or left, or from both flanks) of Companies. { At this caution the rear rank take a pace of nine inches to the front.

Commence Firing. { The flank file at once make ready and come to the present, the front-rank man delivering his fire first, to be immediately followed by that of the rear-rank man; both men then return to the capping position, and go on with their loading in the quick time, performing their motions together and without loss of time. When the flank file is bringing the rifle to the present, the next file is to make ready, coming to the present when the flank file is in the act of returning to the capping position; the next file to proceed likewise, and so continue by files in succession for the first round, after which, each file as soon as loaded fires independently, *i.e.*, without reference to the files either on the right or left.

PLATOON EXERCISE. 93

Cease Firing. { Each file, as it completes its loading, will "shoulder arms." Files that may have made "ready," when this command is given will half cock their rifles and "shoulder arms."

N.B.—Each man, before full cocking his rifle, is to adjust his sight according to the distance at which he estimates the object at which he intends to fire to be. In file and volley firing the front rank must remain perfectly steady after delivering their fire, otherwise the aim of the rear rank will be deranged.

Exercise to receive Cavalry.

The young riflemen, being now presumed to have a thorough knowledge of the preceding portion of the drill, may now be formed into four ranks and practised to receive cavalry, as it is necessary to do, in square four deep.

Prepare to resist Cavalry. } At this caution the second and fourth ranks will take a pace of nine inches to the front.

Ready. { At this command, the first and second rank sink at once upon the right knee as a front and rear rank, kneeling in the manner prescribed when coming to the ready; from shouldered arms, and at the same time place the butts of their rifles on the ground against the inside of their right knees, locks turned up, the muzzle slanting upwards, so that the point of the bayonet will be about the height of a horse's nose: the left hand to have a firm grasp of the rifle immediately above the third band, the right hand holding the small of the butt, the left arm to rest upon the thigh about six inches in rear of the left knee. The third and fourth ranks to make ready as a front and rear rank standing. Muzzles of rifles to be inclined upwards.

Commence Firing from the right (or left, or from both flanks) of Faces.	The standing ranks to commence file firing, in the order before detailed.
Cease Firing.	Each file, as it completes its loading, will shoulder arms.
Kneeling Ranks (or front face, &c., as the case may require), Fire a Volley.	A caution, should it be deemed necessary to fire a volley.
At — yds. Ready.	Come to the capping position, at the same time bring the weight of the body on the right heel, then adjust the sight for the distance named, full cock the piece, and fix the eye STEADFASTLY on an object in front.
Present.	After firing, count a pause of the slow time, and as quickly as possible bring the rifle down to resist cavalry as before directed, remaining perfectly steady.
Load.	Spring to attention at the half-face, and bring the rifle to a horizontal position at the right side, seizing it at the same instant with the right hand close in front of the left, and then come to the position of prepare to load as standing ranks, and go on with the loading in quick time.

NOTE.—In squares of two deep, the front rank *only* kneel to resist cavalry.

MUSKETRY INSTRUCTION.

When our Volunteer has attained a knowledge of the platoon exercise, he is to be instructed in a course of musketry *drill* and *practice* (see "The Rifle, and how to use it," fourth edition, p. 146). To this great object, too much attention cannot be devoted; it is the means by which the soldier is taught *to kill his enemy;* and it cannot be too strongly inculcated that every man who has no defect in his eyes may be made a good shot at a fixed object. The rifle is placed in the soldier's hands for that purpose: his own safety depends on his efficient use of it; and no degree of perfection he may have attained in the other parts of his drill can, upon service, remedy any want of proficiency in this: indeed, all his other instructions can do no more than place him in the best position for using his weapon with effect. The true principles upon which correct shooting may be taught are very simple.

The table given at page 158 of "The Rifle, and how to use it," recapitulates the course of *preliminary training* through which the recruit is to be exercised, before he is to be permitted to fire ball ammunition, also, the *practices in firing ball*, with the number of rounds to be expended in each, which he has to perform before he is allowed to join in the Annual Musketry Instruction of his company.

The table above referred to, enumerates the number of *preliminary drills and practices*, with the number of rounds of ball cartridge to be fired in each, which constitute the course of musketry instruction in which every rifleman should be exercised annually.

Theoretical Principles.

The young rifleman should be made to comprehend the laws which influence the flight of the bullet when discharged, and the reasons for all those rules which have to be attended to in practice; he should therefore make himself thoroughly master—which he can readily do by the aid of this work—of the *construction of the barrel;* of what is to be understood by the "*axis of the piece,*" the "*line of fire,*" the "*trajectory,*" and the "*line of sight;*" how elevation is acquired, and the necessity for it; the importance of holding the sights upright, and the errors which result from inattention to this particular; as also the influence of wind on the bullet when in motion. These several points must be entered into in a clear and concise manner, as it is of importance they should be well understood.

Company Drill.

We may now suppose that the Corps, thoroughly grounded in the rudimentary part of their drill, will be prepared to be taught the movements of the Company, being for this purpose told off in files of from 18 to 20.

It is obviously impracticable, within the narrow limits of this little treatise, nor indeed is it requisite, to go into the details of that drill; we may therefore proceed at once to the consideration of Light Infantry movements.

SECTION VI.

LIGHT INFANTRY.

I.

Light Infantry.—When a regiment is employed as light infantry, it is usually divided into three parts—skirmishers, supports, and reserve; but it may frequently be deemed advisable to cover the movements of a line with skirmishers and supports, or skirmishers only.

II.

Relative Strength of Skirmishers, Supports, and Reserves.—The supports should always be composed of numbers equal to the line of skirmishers; thus, each company, when extended, should have a company to support it. The reserve should be at least one-third part of the whole body.

2. If a single company be detached to skirmish at a distance from the main body, not more than half the men should ever be sent forward to skirmish at a time: the other half remaining formed in support.

III.

1. *Relative Duties of Skirmishers, Supports, and Reserves.*—The movements of the skirmishers depend on the position and movements of the enemy. Care must be taken that the skirmishers protect and overlap the flanks of the main body they are intended to cover.

2. It is the duty of the supports to assist the skirmishers; the movements of the two must therefore correspond. Each support should be, as nearly as possible, in rear of the centre of its own skirmishers. The reserve is the point on which both supports and skirmishers may rally.

3. When the skirmishers are sent out to a distance, the field officers must take care that they are always so situated as to protect the front and flanks of the main body effectually.

IV.

Relative Distances.—The distance of supports from the skirmishers, and reserves from supports, must depend on circumstances. The supports should always be in the most convenient position to assist the skirmishers, without being unnecessarily exposed. Thus, when skirmishers have ascended a bank, and are halted on the summit, the supports may approach closely, without being exposed; but, on a plain, they must be kept at a greater distance. On a plain, the distance between skirmishers and supports should be about 200 yards; between supports and reserves, about 300 yards; between the reserve and main body, 500 yards.

V.

Cover.—1. When under fire, skirmishers must at all times take advantage of all cover, and though not required now to preserve their distances and dressing, they must, when advancing or retiring, take care that they never get in front of each other, and that they never remain under cover so long as to interfere with the fire of their comrades.

2. Officers commanding supports, must take advantage of all objects affording cover to protect their men, and should make them lie down when cover can be obtained by so doing.

3. The officer commanding the reserve must also keep his men under cover when practicable.

4. In presence of cavalry, the reserve must keep in column; but under the fire of artillery, it should be deployed into line.

VI.

Time of Movement.—Light Infantry movements are usually performed in quick time, except extensions or closings on the march, the formation of company or rallying squares, and changes of front from the halt, which will be in double time. When more than usual rapidity is required, the men may be directed to "double."

VII.

Points of Direction.—All lines of skirmishers move by their centre, except when inclining to a flank, in which case they move by the flank to which they are inclining.

VIII.

How Arms are Carried.—The skirmishers and supports move with trailed arms, except in close column of sections, or in files,

LIGHT INFANTRY.

when they move with shouldered arms; reserves move with sloped arms.

IX.

Officers and Connecting Links.—1. When a company is extended, the captain should be at a convenient distance in rear of the centre; the supernumeraries being placed at equal distances along the rear of the line of skirmishers, the lieutenant being always near the outer flank of a flank company. When a company is in support, the captain should be in its proper front, whether it is advancing or retiring, he will thus lead his company when advancing, and follow it when retiring. The supernumeraries of a support will be in the rear. The officers of a reserve will always be placed as in column, right in front.

2. A non-commissioned officer, or two, may frequently be sent out with advantage from a support to keep up the connexion with its skirmishers; these men are termed connecting links.

X.

Words of Command and Bugle Sounds.—1. Light infantry movements should, as a rule, be regulated by word of command. Commands being repeated by the captains and every supernumerary belonging to the line of skirmishers. The connecting links may be employed, when necessary, to pass words of command, or convey intelligence between the reserve and supports, &c. When, on account of the distance, or from noise or wind, the voice cannot be distinctly heard, the connecting links should run up, deliver their orders, and resume their places.

2. Calls on the bugle may occasionally be necessary as substitutes for the voice, but as they are constantly liable to be misunderstood, and as they reveal intended movements to the enemy, who soon become acquainted with them, they should seldom be used, unless for purposes of drill.

3. Bugle sounds must be as few and as simple as possible. None but the following sounds ought ever to be used in light drill:

4. One G denotes the right of the line. Two G's the centre. Three G's the left.

5. The G's preceding any sound denote the part of the line to which it applies.

I. EXTEND.

100   RIFLE VOLUNTEERS.

The Halt annuls all previous Sounds except the Fire.

This sound is used to turn out troops in cases of alarm. It must never be used for any purpose but that above stated.

LIGHT INFANTRY.   101

IX. INCLINE.

X. WHEEL.

The calls IX and X are of course preceded by the distinguishing G's.

XI. THE ALARM, OR LOOK OUT FOR CAVALRY.

XII. THE OFFICERS' CALL.

XIII. THE QUICK TIME.

XIV. THE DOUBLE TIME.

6. Every regiment ought to have a well marked and simple regimental call of its own.

7. The Advance or the Retire sounded when inclining to the flank, indicates that the original direction is to be resumed.

8. When moving by sound of bugle, the men are always to wait till the bugle has ceased before they move.

9. When THE FIRE is combined with any other call, it should always be the last sounded, for if the men commenced firing they would not hear the second call.

10. Bugle sounds never apply to bodies of troops in reserve.

LIGHT INFANTRY MOVEMENTS OF A COMPANY.

The following general rules have been laid down for the guidance of Light Infantry, but all movements in extended order depend so much upon circumstances, that officers must rely to a great extent on their own judgment for the effectual performance of their duties.

Volunteers should first be instructed in the following movements on level ground, keeping their distances and dressing in extended order; when they have learnt these, they must be taught to apply them practically.

1. *Extensions.*

In extending, the rear rank man of each file should regulate the distance, and the front man should look to the direction.

The number of paces that files are to extend from each other may be specified in the caution thus:—FOUR PACES FROM THE RIGHT—EXTEND. When no number is specified, six paces is the regulated distance between files.

From the Right (Left, Centre, or No. —, File)— Extend. Bugle Call, No. I.

1. *From the Halt.*—On the word EXTEND, the captain places himself in rear of the centre of the company, the senior supernumerary in rear of the right, and the second senior in rear of the left. The file on the named flank, or the centre or named file, kneels down, the remainder shoulder arms, face outwards, and extend in quick time. The front rank men move direct to the flank, covering correctly on the march, the rear rank men will cast their eyes over the inward shoulder, and tap their respective front rank men as a signal to halt, front, and kneel, when they have gained their proper distances.

The men must learn to extend from any file of a close column of sections, without previously re-forming company; the named file will kneel, and the remainder will face outwards and extend as already described.

LIGHT INFANTRY. 103

From the   2. *On the March.*—On the word EXTEND, the
Right (Left, file on the named flank, or the centre or named file,
Centre, or continues to move straight forward in quick time,
No. —, the remainder make a half turn to the flank to
File.)— which they are ordered to extend, and move off in
Extend. double time. As soon as each file has extended to
Bugle Call, its proper distance, it turns to its front and resumes
No. I. the quick time; the rear rank men covering their
front rank men, and the whole keeping in line with the directing
file.

Men in extended order may be directed to increase the distance between their files any given number of paces, from either flank, the centre, or any named file. If the bugle sound the EXTEND, the skirmishers will open out one half more than their original extension; thus, if they are at six paces' distance, they will open to nine.

When a company, extending on the march, halts before all the files are extended, the remainder make a half turn outwards into file, break into quick time, advance arms, and complete their extension as from the halt.

2. *Closing.*

On the   1. *On the Halt.*—On the word CLOSE, the file on
Right (Left, the named flank, or the centre or named file, will
Centre, or rise, order arms, and stand at ease; the remainder
No. —, rise, face towards it, and close at quick time, halt-
File)— ing, fronting, ordering arms, and standing at ease
Close. in succession as they arrive at their places; the
Bugle Call, officers remain in the rear unless directed to take
No. II. post.

The file on which the skirmishers close may be faced in any direction; the remainder will form upon it, facing in the same direction.

On the   2. *On the March.*—On the word CLOSE, the file
Right (Left, on the named flank or centre, or the named file,
Centre, or will move steadily on in quick time; the remainder
No. —, File) will make a half turn towards it and close at the
—*Close.* double, turning to the front, and resuming the
Bugle Call, quick time as they arrive at their places.
No. II.

When a company, closing on the march, is halted before all the files are closed, the remainder make a half turn inwards into file, break into quick time, and complete the formation as from the halt.

3. Squares.

Company Squares.

In light infantry movements, company squares will be formed thus:—

Form Close Column of Sections. On the word SECTIONS, the right section face to the left, and disengage to the front by the leading file closing two paces to the right, the front rank man inclining rather back; the third and fourth sections will face to the right and disengage to the rear in the same manner.

The men, when forming from close order, move into column with their arms shouldered, the second section fixing bayonets on the word *March*, the remaining sections as they halt and front, and when they run in from extended order they order arms and fix bayonets independently as they halt and front in their places.

Quick-March. On the word MARCH, they step off and form close column on the second section, halting and fronting without word of command as they arrive in column: the distance between the sections being one pace; the captain places himself on the left of the front rank of the leading section, covered by his covering serjeant, the supernumeraries on the reverse flank of their respective sections.

Prepare for Cavalry. On the words PREPARE FOR CAVALRY, the officers and non-commissioned officers move into the centre of the column; the men then face outwards, so as to show a front of equal strength in every direction.

Ready. On the word READY, if the square is two or three deep the front rank only will kneel; if four deep, the two front ranks kneel. The remainder will come to the Ready.

The Rallying Square.

Form Rallying Square. The men being dispersed, on the words FORM RALLYING SQUARE, an officer, as a rallying point, holds up his sword and faces the supposed enemy; the men hasten to him, fixing swords and ordering arms as they reach him. The two first form on his right and left, facing outwards. The three next place themselves in front of those posted, and three others in rear, facing to the rear, thus forming a square. The next four men take post at the several angles; and others as they come up will complete the different faces between these angles.

4. *Advancing in Skirmishing Order.*

Company Advance.
Bugle Call, No. III.

On the word ADVANCE, the men rise and step off in quick time with trailed arms, keeping their distance from the centre.

5. *Retiring in Skirmishing Order.*

Company Retire.
Bugle Call, No. VII.

On the word RETIRE, the men rise, face to the right about, and step off in quick time, rear rank in front, keeping their distance from the centre.

Men in extended order will invariably face or turn to the right about, whether advancing, retiring, firing, or not firing.

6. *Passing Obstacles in Skirmishing Order.*

Men in extended order must frequently be practised in passing obstacles both in advancing and retiring. When an obstacle, such as a pond, presents itself in front of a line of skirmishers, the files opposite to it open out gradually as they approach, and pass on either side, closing upon the remaining files, which continue moving straight to their front. Having passed the obstacle, the files that have been diverted again extend and fill up the interval in the line.

7. *Inclining to a Flank.*

To the Right (or Left)— incline.
Bugle Call, one G (or three G's) followed by No. IX.

On the word INCLINE, the skirmishers make a half turn to the flank to which they are ordered to incline, and move diagonally till they are ordered to resume their original direction, by the word " Advance" or " Retire." If the skirmishers have made a half turn, and are again ordered to incline in the same direction, they complete the turn by making a second half turn, and take ground to the flank and file.

If the halt sounds when men are inclining, they halt, front, and kneel.

8. *Skirmishers changing Front or Direction from the Halt.*

A line of skirmishers halted, can change front on any two named files, placed as a base, and at any angle.

Change Front to the Right (or Left) on the two Centre (or on No. —, and No.—) Files.

1. *From the Halt.*—On the caution,—the two named files rise. The captain dresses them in the direction required; as soon as they are placed, they kneel again.

Double March.

On the word MARCH, the whole will rise, and if all the files are to be thrown forward on a flank, they make a half-face inwards, and move across by the shortest way to their places in the new line, dressing on the two base files, as they successively halt, and then kneeling.

The young volunteers should first be taught this movement in quick time, and by separate words.

Skirmishers (Right or Left) Wheel. Bugle Call, one G (or three G's), followed by No. x.

2. *On the March.*—A line of skirmishers on the march may change their direction gradually, as a company wheels. On the word WHEEL, the pivot file halts, and the remainder circle round it, the front rank men looking outwards for the dressing, and the rear rank men keeping the distances from the pivot flank.

Forward. On the word FORWARD the whole line advances by the centre.

9. *Firing in Skirmishing Order.*

The men of a file must always work together. Both men should never be unloaded at the same time; they should always load when practicable under cover; before moving to the front, when advancing, and after falling back, when retiring.

Commence Firing. Bugle Sound, No. v.

1. *Firing on the Halt.*—On the words COMMENCE FIRING, the front rank men make ready, fire, and load; the rear rank men, when their front rank men are in the act of capping, make ready, fire, and then load.

Skirmishers may be ordered to lie down, for the sake of cover. When firing thus, both elbows rest on the ground to support the body and rifle; the men loading on their knees. Riflemen may fire on their backs in favourable situations; the feet are then to be crossed, the right foot passed through the sling of the rifle, and the piece supported by it. In very exposed situations, if a rifleman wish to load lying, he will roll over on his back, and place the butt of his firelock between his legs, the lock upwards, and the muzzle a little elevated.

Commence Firing.
Bugle Sound, No. v.

2. *Firing when Advancing.*—On the words COMMENCE FIRING, the whole of the skirmishers will make a momentary halt, the front rank man of each file fires (kneeling if preferred), and takes a side pace to his left; the rear rank man then passes on, the front rank man following close behind him, loading on the march; when in the act of capping he gives the word "Ready" in an under tone, on which the proper rear rank man fires, and both men will proceed as above described.

When men find difficulty in loading on the march, they may halt and load, and then double up to their file leaders.

When cover presents itself, the men must be taught to take advantage of it; when any large object comes in their way, several files may run up behind it, fire, load, and then move on and regain their distances and places in the general line.

Commence Firing.
Bugle Sound, No. v.

3. *Firing when Retiring.*—On the words COMMENCE FIRING, both ranks halt and front, the front rank man of each file fires, faces to the right about, and retires in quick time, passing by the left of his rear rank man (who follows close behind), loading as he retires; when his loading is completed, both ranks halt and front, the rear rank man firing and proceeding in the manner described for the front rank man.

On rough ground, files will run back from one cover to another, taking care before they leave one to select another. One man of each file should fire previous to moving, and re-load when again under cover. As the object is to keep the enemy in check, skirmishers when retiring should hold each station as long as possible without risk of being cut off by the enemy, or of being shot by their comrades.

When a line of skirmishers halted, is ordered to advance or retire firing, the front rank men first fire, the whole then rise, and proceed as already described.

4. *Firing while inclining to a Flank, or taking Ground to a Flank in Files.*—When skirmishers are ordered to fire, while they are inclining to the right or left, or taking ground to a flank in files, the front rank men will halt, take steady aim and fire,

the rear rank men moving on; having fired, the front rank men will double up to the proper rear of their rear rank men, and then load on the march, or load at the halt, and then double up. When their loading is completed, the rear rank men proceed in like manner.

When skirmishers, either halted or on the march, are directed to CEASE FIRING, they are to complete their loading, and the rear rank men are to resume their places in the proper rear of their front rank men, if not there already.

Whenever skirmishers are directed to halt, by the word of command or bugle sound, they will halt and kneel, facing to their proper front, and if firing, they must continue firing.

SECTION VII.

MISCELLANEOUS MOVEMENTS.

Before bringing this little treatise to a close, I thought it might be advisable to introduce a few simple battalion movements, in order to give a general idea of some of the more ordinary infantry evolutions of this kind in the field. To render them readily comprehensible, they are illustrated by diagrams.

A battalion in extended order retires across a bridge, in contact with an enemy, in the following manner:

1. *Advancing.*—The skirmishers, on reaching the margin of the river, lie down and cover themselves, keeping up a brisk fire

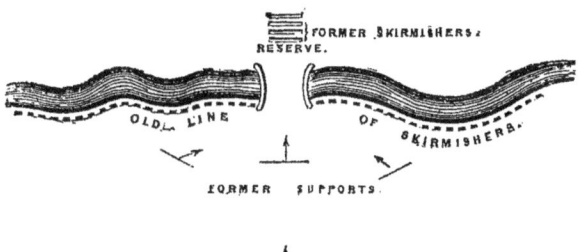

upon the enemy. The supports, on approaching the bridge, close upon the opposite support, and aided by the reserve, force the passage with the bayonet; having succeeded in this, and the reserve having crossed the bridge, the supports extend from their centre, the reserve holding the bridge, the first skirmishers keeping up a heavy fire, until clouded in succession by the new line. The new line completes its extension; the reserve sends out fresh supports, and the old skirmishers assemble in rear of the reserve. The whole, then move forward according to the original formation.

2. *Retiring.*—The reserve pass first, and take post at the bridge-head; detaching parties to both flanks in extended order to line the river. The supports close upon the one opposite the bridge, and in compact order, halt in front of it, till the line of skirmishers is withdrawn. That this may be done rapidly, the skirmishers inclining towards the bridge when at some distance, and on arriving near it, run briskly across, forming in rear of the reserve. The supports then cross, and join the reserve, the whole being prepared to defend the bridge or retire, as may be ordered. The new line of skirmishers commence firing as soon as their front is clear; and if the retreat is to be continued, supports between them and the reserve are again thrown out. The mode in which these movements are effected is represented in the annexed woodcut.

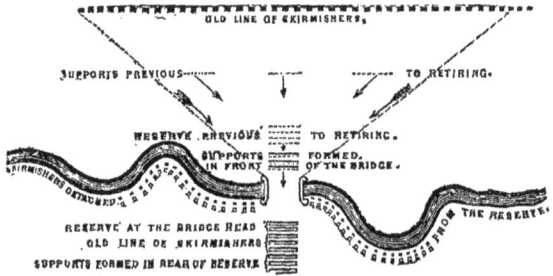

A Battalion extending in Skirmishing Order from Line.

When a battalion extends from line, the skirmishers may be taken from either flank; the companies next to them form the supports, the remaining companies the reserve. The extension is thus performed:—

A Battalion of Ten Companies extending from Line, Three Companies skirmishing.

Word of Command.—The Battalion will skirmish, three companies on the right —*From the Centre Extend.*

Movements of Right Skirmishers.—No. 1. *Double March, By Sections Right-Wheel, Forward* (in échellon), as soon as the centre skirmishers are extended—*From the Left*—Extend.

Movements of Centre Skirmishers.—No.2. *Quick-March, from the centre,* Extend.

Movements of Left Skirmishers.—No. 3. *Double March, By Sections, Left-Wheel, Forward* (in échellon), as soon as the centre skirmishers are extended—*From the Right Extend.*

Movements of Right Support.—No. 4. *Quick-March, By Sections Right-Wheel, Forward* (in échellon), when in rear of the centre of the right skirmishers, *Reform-Company.*

Movements of Centre Support.—No. 5. *Quick-March. By Sections Right-Wheel, Forward* (in échellon), when in rear of the centre of the centre skirmishers, *Reform-Company.*

Movements of Left Support.—No. 6. *Quick-March*, and move in rear of the centre of the left skirmishers.

Movements of Reserve.—Nos. 7, 8, 9, and 10 Companies will form the reserve, in line or column, and will move in rear of the centre by fours.

A battalion of six companies extends from line in the same manner as one of ten, the two companies on the right (or left) skirmishing, the two next in support, the remainder in reserve.

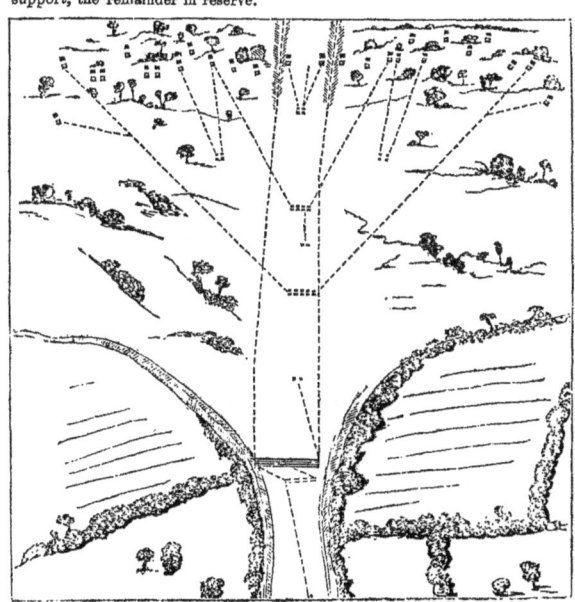

112 RIFLE VOLUNTEERS.

The Formation of an Advanced Guard on a Plain is thus effected :—

An advanced guard on a plain is simply a line of skirmishers with supports, and if far distant from the main body, with a reserve. When the leading files come out on a plain they halt and lie down, the flanking files moving up and lying down in line with them at the usual distance of 100 yards. The remainder of the leading section or subdivision as it comes up, extends from its centre; the second section or subdivision also extends from its centre, and reinforces the leading section or subdivision. The skirmishers thus formed advance, correcting their distances from the centre on the march; the reserve form subdivision or company, and support the skirmishers. This formation is very useful when the leading files have discovered an enemy without having been themselves observed.

An advanced guard, thus extended, may resume its original formation by the leading files moving on, and the remainder halting till they have gained their proper distances, and then following on in succession, the remainder of the first and second sections or subdivisions closing on their centres.

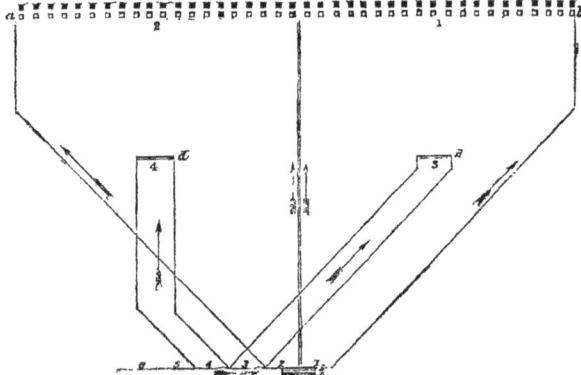

A SHORT ABSTRACT

OF THE

Laws at present in force relating to Volunteer Corps, whether Infantry or Cavalry.

Acts previous to 44 Geo. III. c. 54, repealed by that Act so far as they refer to Yeomanry and Volunteer Corps.

The Acts having reference to VOLUNTEER and YEOMANRY CORPS were consolidated by the 44th Geo. III. c. 54. The following are the principal provisions of this Act, which is at present in force:—

By § 3, Her Majesty may continue the services of all corps of Yeomanry or Volunteers accepted before the passing of that Act (5th June, 1804), and may also accept the services of any corps of Yeomanry or Volunteers that may be formed after the passing thereof, such corps respectively being formed under officers having, or who shall have, commissions either from her Majesty or any Lieutenant of a county, or any other person or persons who may be specially authorized by her Majesty for that purpose, upon such terms and conditions, and under such regulations as have been or shall be approved by her Majesty in regard to such corps. And her Majesty may disband or discontinue the services of any such corps, or any portions of such corps, whenever it may seem expedient to her Majesty to do so. Provided always that the services of all corps of Yeomanry and Volunteers accepted before the passing of that Act shall be deemed to be continued under the provisions thereof, unless her Majesty shall signify her intention of disbanding or discontinuing the services of any such corps by any order to be communicated by her Majesty's principal Secretary of State.

By § 4, Effective members of Yeomanry or Volunteer corps are exempted from service in the Militia or other additional forces, except in the case of corps whose offers of service specify that no such exemption would be claimed, and no such exemption is to extend to any greater number than the established number of such corps.

§ 5. Only those are to be deemed effective members who have attended muster or exercise, properly armed and accoutred, if cavalry, four days, if infantry, eight days at the least in the four months immediately preceding the return required by the Act, and who have been duly returned by the commanding officer as effective members, and as having taken the oath of allegiance.

§ 6, Contains an exception when there has been a delay in supplying arms.

§ 7. The commanding officer may grant leave of absence, and such absence shall not prevent the member so obtaining it from being returned as effective, provided that during the next four months he serves as many days as shall make up for the whole period of eight months, if he is in the cavalry, eight—if in the infantry, sixteen—days' exercise. If he does not complete this, he is to be struck off the list of effective members, and to be returned in the muster-roll as non-effective.

§ 17.* Persons so returned as effective may be balloted for the Militia, and immediately on their ceasing to be returned as effective, they are liable to serve.

§ 8.† It is sufficient, to render a man effective, that he attend during the whole year, if in the cavalry, twelve—if in the infantry, twenty-four—days within one period, or two successive periods of four months next before the return.

§ 9.‡ Commanding officers are to make a return on the first day of every April, August, and December to the clerks of lieutenancy of the numbers of men in the corps, and of the number of supernumeraries, distinguishing between the effective and the non-effective members, of the persons who have entered the corps since the last return, of those who have been absent on leave, and of those who have been discharged from or have quitted the corps since the last return; and where any arms have been required by such corps at the expense of her Majesty, and have not been supplied, such circumstance is to be stated at the foot of the return. The commanding officer is also to send in to her Majesty's principal Secretary of State, and to the general officer commanding the district, if any, accurate returns of the effective and non effective men in the form of the usual military returns.

§ 10. Commanding officers are required to give certificates to effective men residing in other places, which shall entitle them to exemptions therein.

§ 11. Field officers and adjutants of Volunteer corps, and persons serving in Yeomanry or Volunteer cavalry, are exempted from duty for horses used at muster and exercise, and also persons providing them; and all effective members of Yeomanry or Volunteer corps from the hair-powder duty.

§ 12. No corps is to be entitled to exemptions unless the com-

* Amended by 53 Geo. III. c. 84, § 4.
† Amended to 6 days in the year, 2 days in each 4 months, or 5 successive days, 56 Geo. III. c. 39, § 1.
‡ The returns are to be sent in once a year, within 14 days of Aug. 1, 7 Geo. III. c. 58, § 2.

manding officer certify in the muster-rolls that it has been, or has been ready to be, inspected.

§ 13. No toll is to be demanded for any horse ridden by any person in any corps of Yeomanry, or by any field or staff-officer of Volunteers going to exercise, &c., dressed in uniform, and armed and accoutred.

§ 15. Commanding officers making false returns, or giving false certificates, are subjected to a penalty of 200*l.* for every offence.

§ 20. Every person enrolled is to take the oath* of allegiance, which may be administered by any deputy-lieutenant, justice of the peace, or commissioned officer of the corps.

§ 21. Adjutants, serjeant-majors, and others who are receiving constant pay, are subject to the Mutiny Act and to the Articles of War; every court-martial in such case is to be composed wholly of members taken from the Yeomanry or Volunteer establishment, and no punishment is to extend to life or limb, except when the corps is called out in case of an invasion.

§ 22. In all cases of actual invasion, or appearance of an enemy in force on the coast of Great Britain, or of rebellion or insurrection arising or existing within the same, all corps of Yeomanry or Volunteers shall, whenever they shall be summoned by the lieutenants of the counties in which they shall be respectively formed, or their vice-lieutenants or deputy-lieutenants, or upon the making of any general signals of alarm, forthwith assemble within their respective districts, and shall be liable to march according to the terms and conditions of their respective services, whether the same shall extend to any part of Great Britain, or be limited to any district, county, city, town, or place therein; and all persons then enrolled in any such corps, not labouring under any infirmity incapacitating them from military service, and not holding a commission or serving in any of her Majesty's other forces, or in any other such corps of Yeomanry or Volunteers, and actually joining such corps, who shall refuse or neglect to join their respective corps, and to assemble and march therewith, upon any such summons or general signal of alarm as aforesaid, shall be deemed deserters, and shall be subject to punishment as such; and all such corps of Yeomanry or Volunteers, and all officers and non-commissioned officers, drummers, and private men therein, shall, upon and from the time of such summons or of such general signals of alarm being made as aforesaid, and until the enemy shall be defeated and expelled, and all rebellion or insurrection then existing within Great Britain shall be suppressed (the same to be signified by her Majesty's proclamation), continue and be subject to

---

* The following is the form of oath:—"I, A— B—, do make oath, that I will be faithful and bear true allegiance to her Majesty, her heirs and successors, and that I will, as in duty bound, honestly and faithfully defend her Majesty, her heirs and successors, in person, crown, and dignity against all enemies, and will observe and obey all orders of her Majesty, her heirs and successors, and of the generals and officers set over me. So help me God.

"Sworn before me, C— D—, this — day of —, 1859."

all the provisions contained in any Act of Parliament then in force for the punishment of mutiny and desertion, and for the better payment of the army and their quarters, and to any Articles of War made in pursuance thereof in all cases whatever.

§ 23. Whenever any corps of Yeomanry or Volunteers shall, with the approbation of her Majesty, signified through her principal Secretary of State, voluntarily assemble or march to do military duty upon any appearance of invasion, or for the purpose of improving themselves in military exercise, except in the case hereinafter specified as to corps of yeomanry cavalry, or shall voluntarily march on being called upon in pursuance of any order from the lieutenant or sheriff of the county, to act within the county or adjacent counties for the suppression of riots or tumults, all such corps of Yeomanry or Volunteers shall in all such cases, from the time of so assembling or marching as aforesaid, and during the period of their remaining on such military duty, or being engaged in such service as aforesaid, be subjected to military discipline and to all the provisions of any Act then in force for the punishment of mutiny and desertion, and for the better payment of the army, and their quarters, and to any Articles of War made in pursuance thereof.

§ 24. Her Majesty may put such corps under the command of such general officer as she shall appoint; but such corps shall be led by their respective officers, and no effective member shall be liable to be placed in any other regiment.

§ 25. No officer of Volunteers is to sit on the trial of any officer or soldier of the other forces, and contrariwise.

§ 26. All officers in corps of Yeomanry or Volunteers having commissions from her Majesty, or lieutenants of counties, or others who may be specially authorized by her Majesty for that purpose, shall rank with the officers of her Majesty's Regular and Militia forces, as the youngest of their respective ranks.

§ 27. Commanding officers of Yeomanry or Volunteer corps, when not on actual service, may discharge members, not being commissioned officers, for disobedience of orders, &c.

§ 28. When the regulations of a corps do not provide for any case of misconduct under arms, the commanding officer may disallow the day on which the party misconducted himself as a day of attendance.

§ 29. Persons misconducting themselves during exercise may be ordered into custody for the time during which the corps remains under arms.

§§ 30 and 31. Persons enrolled as Volunteers may quit their corps, except when called out in cases of invasion, &c., except the persons receiving the constant pay of their rank. None can quit, however, without notice of their intention to quit, nor till their arms, &c., shall have been delivered up, and all fines paid, unless by enlisting in her Majesty's forces or being enrolled in the Militia.

§ 33. Persons thinking themselves aggrieved by the commanding officer refusing to strike their names out of the muster-rolls, may

VOLUNTEER CORPS.

appeal to two deputy-lieutenants, or one and a justice, who may determine the same.

§ 36. When Volunteers are assembled on summons of the county lieutenant, &c., or on a general signal of alarm, the receiver-general of the duties under the commissioners for taxes in England, and the collector of the cess in Scotland, are to pay to the captain of the troop or company two guineas for the use of every Volunteer in such troop or company who shall so assemble, and, when voluntarily assembled, the Treasury may order a guinea for each to be paid in like manner. The captains are to account to the men for money, and not to draw any for the use of men not desiring it.

§§ 37 and 38. Volunteers, when assembled on invasion, &c., are entitled to receive pay, and to be billeted as other forces, and their families are entitled to the same relief as the families of Militiamen.

§ 39. After the defeat and expulsion of the enemy, and after the suppression of any rebellion or insurrection, the Volunteers are to be returned to their respective counties, and a guinea paid to each man willing to receive it.

§ 40. Commissioned officers disabled in service, are entitled to half-pay, and non-commissioned officers and privates to Chelsea Hospital; and widows of officers killed in service to pensions for life.

§ 41.* Half-pay may be received by adjutants and quartermasters on taking the oath that they have not any place or employment of profit, civil or military, under her Majesty.

§ 42.† Commanding officers may appoint places for depositing arms and accoutrements, and persons to take care of them; and the deputy-lieutenants shall view them; and the expense shall be paid in England by the receiver-general of the county.

§ 44. In case any man shall sell, pawn, or lose any arms, accoutrements, clothing, or ammunition delivered to him, or shall wilfully damage any such arms or accoutrements, every such man shall, for every such offence, forfeit and pay a sum not exceeding forty shillings, and if not paid, the party may be committed.

§ 46. When corps of cavalry shall be desirous of assembling under the command of their own officers, the county-lieutenant, with the approbation of her Majesty, may make an order for that purpose, and an order to any justice of the county, who shall issue his precept for billeting the non-commissioned officers and privates as her Majesty's forces may be billeted; but corps so assembled shall not be subjected to the mutiny laws.

§ 47. The Acts for billeting her Majesty's forces extend to such corps when billeted.

§ 48. When the lieutenant has fixed the day and place of exercise for Yeomanry or Volunteer corps, he is to certify the same to the Secretary-at-War.

---

* An officer on half-pay, however, does not forfeit it by reason of his holding a commission in the Yeomanry, and receiving pay as such, 57 Geo. III, c, 44, § 2.
† By the Secretary of War, 7 Geo. II. c, 58, § 4.

118 LAWS RELATING TO VOLUNTEER CORPS.

§ 50. The property in subscriptions, arms, &c., is vested in the commanding officer for all purposes of indictments or suits.

§ 51. If subscriptions or fines be not paid, a justice of the peace may direct double the amount to be paid, which may be levied by distress.

§ 56. No future rules or regulations are to be valid or binding on any corps of Yeomanry or Volunteers, unless submitted to the principal Secretary of State, and not disallowed by her Majesty.

§ 58. The acceptance of a commission in any corps of Yeomanry or Volunteers does not vacate a seat in Parliament.

§ 60. Provisions relating to corps are to extend to independent troops or companies.

THE END.

www.ingramcontent.com/pod-product-compliance
Lightning Source LLC
Chambersburg PA
CBHW051808040426
42446CB00007B/579